1981

University of St. Francis
GEN 791.4501 B144
Baggaley, Jon.
W9-ADI-741
3 0301 00084273 8

PSYCHOLOGY OF THE TV IMAGE

Psychology of the TV Image

JON BAGGALEY
MARGARET FERGUSON
PHILIP BROOKS

PRAEGER SPECIAL STUDIES • PRAEGER SCIENTIFIC

Published in 1980 by Praeger Publishers
CBS Educational and Professional Publishing
A Division of CBS, Inc.
521 Fifth Avenue, New York, New York 10017 U.S.A.

Library of Congress Catalog Card Number: 79-92118

0123456789-056-987654321
Printed in Great Britain
ISBN: 0-03-046206-1

Contents

Tables

Figures

NB The following Appendices may be obtained from the author:

1 Experimental texts used

2 'British Vagrancy Trust' material

3 Semantic differential scales used

4 Factor solutions: Experiment 19

Preface

Readers must have a reason for studying a book in depth. As the range of material in a given field proliferates, the reasons for concentrating on any one piece of work become increasingly arbitrary. Even a commitment to its subject matter may no longer govern a reader's decision to select the work and follow it through. Before turning to the first page of the Preface he rapidly scans the pages that follow, lining up expectations regarding the work as a whole. If he is worldly wise he checks on its author and on any claims by the book to serious attention. Though he might accept it only reluctantly, the style of print, cover design and complexity of diagrams can each affect his attitude and approach to the reading task at hand. Or curtail it abruptly. For time and education have taught us to be critical of the books and people that try to tell us things, and to invest our efforts in them with care.

So it is when we meet people face to face; and so it is also when they confront us through the medium of television. As the most pervasive of modern technological stimuli TV offers greater satisfaction to more appetites than any other medium yet devised, via more alternative modes of expression than we can comfortably deal with. It is possible to resolve this dilemma by becoming connoisseurs of its surface qualities, choosing between news broadcasts on the basis of their stylistics and registering the visual impact of an advert more adeptly than the article it promotes. As one becomes responsive to the medium's artistry its messages become remote; pre-tuning our attention upon them becomes a matter of conscious effort.

In the pages that follow we consider the effects of TV in this light; not merely as a vehicle for opinion and fact but as a source of artistic images with possible meanings in their own right. We refer, however, to the artistry employed not for art's sake, when the medium is used for entertainment purposes, but to the aesthetic effects produced in the service of news and current affairs, commercial information and education. All communication depends upon skills of presentation, and when it occurs via TV an extensive repertoire of techniques is drawn upon. They produce a range of 'meta-messages' intended to convey the main content as effectively as possible. But at times they defeat the intended message with an impact of their own; while at others they bolster it to a persuasive status it may not deserve.

For the TV producer as well as the viewer, therefore, an understanding of the effects of TV presentation can provide a finer control over the advantages it can bring and safeguards against abuse. We report a series of investigations designed to increase this awareness, examining the effects due to production facets such as camera angle, visual background detail and editing technique, studio seating position, dubbed music, subtitles, autocue and styles of performance. Each of these involves a range of options taken in TV production for a commonly accepted range of reasons, some artistic, some expedient. Each contributes to the imagery received by the viewer, and may thereby affect his responses. But are audience responses to TV content necessarily as the producers intend? In designing these experiments our assumption was that they may not be: that the imagery of TV can unwittingly subvert its intended effects with uncontrolled consequences.

The book develops a research theme discussed in an earlier title in this series, *Dynamics of Television* by Baggaley and Duck (1976). The main perspective on that occasion was theoretical–linguistic and psychological notions about communication being applied in the study of one of its media. On this occasion we examine empirical evidence for the same thesis, submitting a variety of patient subjects to a range of presentation styles and comparing their reactions. In the early sections the theoretical position taken in the previous book is paraphrased, and the methodological basis discussed for the research which follows. The thirty experiments reported build upon and extend the previous work, relating the effects of verbal content to that of images couched in other aural and visual forms. In referring to the TV image as such, therefore, we do not imply that the study of its visual effects exhausts the matter. Though the visual images are the most obvious of TV's expressive forms, presentation of, for instance, musical imagery in parallel can create psychological effects deserving much future attention. To understand these we must study the mental images TV conveys to the individual, and the public image acquired by a performer when different viewers' attitudes towards him agree. Our use of the term is thus intentionally ambiguous, implying each and all of these meanings as aspects of the one psychological problem.

We believe that the general weight of evidence reported justifies us in identifying the question of 'non-verbal mediation' as a substantive one for study. Occupying the same position within media studies as occupied by the processes of non-verbal communication in studies of social interaction generally, non-verbal mediation processes are based on a unique brand of logic arising from the distortions to a message introduced by mediating technique. The purpose of experimental study is to pinpoint these distortions and to establish their psychological importance in different situations.

Can the imagery of TV be put to new and more systematic uses on this basis? Its psychological effects, though complex, are usually quite logical, and the scope for TV producers to develop fresh conventions harnessing them is vast. In concluding the book we suggest a set of guidelines for the TV producer and viewer wishing to increase his critical control over the medium's persuasive powers. Such guidelines, like the mosaic of effects offered by the book as a whole, can never be complete, for however thoroughly we scrutinise them human reactions will always remain on the move. We do not attempt in this work, therefore, to pinpoint the psychology of the TV image in hard and fast terms for the indefinite future, merely to indicate the subtle variety of its levels and ways in which its facets may be examined in different situations. Rather as the early investigators of personality and intelligence, we seek to extend our chosen field of study by the development and application of reliable methods of measurement.

We have continued to adopt one practice in the book already made clear in *Dynamics of Television.* When other areas of media research have to our knowledge been adequately summarised elsewhere we do not repeat the effort. Media theory now contains numerous sub-sections, superimposed, in parallel, some with political motivations, others as far as they can be purely artistic in their intentions. Few are actually incompatible, for the problems they tackle are highly diverse. When our ideas have been influenced or challenged by other approaches we hope to have acknowledged them. If, however, another approach merely shares a similar tradition to our own it is the tradition that we shall probably stress in our discussions, rather than embroiling the whole with divers points concerning quite different problem areas. Our purpose is the habilitation of non-verbal mediation and its psychological impact within the communication studies mainstream, rather than a celebration of the latter *per se.*

Several elegant research approaches fall outside the confines of these criteria–aptitude treatment interaction studies, the 'uses and gratifications' approach, the information processing models. We trust that our decision to emphasise other paradigms does not appear inflamatory to those who have been diligent in these areas; and we hope that our necessary inclusion of numerous empirical and analytic arguments does not prove too taxing on the general reader unversed in experimental methodology.

We extend our gratitude to the Social Science Research Council for its support to the work between 1976 and 1978. The work has been sustained from the outset by the sound advice and encouragement of Dr Steve Duck: a substantial number of the ideas in the book flow directly from his influence. Our other major debt is to Dr Harry Jamieson for providing the atmosphere of constant stimulation that makes ideas grow and develop. For valuable assistance at various stages of the research we thank John Robinson, Ian Woolf and Neil Barnes of the BBC Education Department; John Caine of BBC Radio Merseyside; David Hyde and the technical staff of our department at Liverpool University; Liverpool Education Authority; Gary Coldevin, Arthur Sullivan and our other Canadian colleagues–including John Thompson; also Renee and Steve Filbin for help with artwork and computer programming We acknowledge with warmth the friendly encouragement and posthumous memory of Richard Sherrington. To the other members of our 'team', Margaret Johnson and Kristina Spencer, we owe an award for patience and endurance. We are particularly grateful to Sheila Baggaley for her work on the manuscript and bibliography; and we acknowledge the self-effaciveness of all those who have allowed us to edit, manipulate, scrutinise and factor analyse them on videotape. Finally, we thank the teachers, pupils, secretaries, clerical staff and undergraduates who have viewed for us through 897 experimental sittings–and Sheila, Jan, David, Helen and Edward for standing for it!

Few investigations ever innovate or clinch a problem: it has always been sewn up beforehand by either Shakespeare, Oscar Wilde or Cole Porter. The time worn maxim underlying our own research is: 'It's not what you say but the way that you say it'. For the sake of those who make TV and those it touches we hope that our illustrations in this respect will help to draw out the question of media technique for attention in further contexts.

JPB
Liverpool University
May 1979

Acknowledgements

Experiments 1 to 6 (Chapter 2) and 18 are abbreviated reports of work discussed by the author and Dr Steven Duck in *Aspects of Educational Technology 8* (pub. Pitman, 1975); *Dynamics of Television* (Saxon House, 1976); *Educational Broadcasting International* (British Council, 1974-75); *Journal of Educational Television* (Educ. TV Association, 1979). Chapter 1 contains material also published in the *Journal of Educational Television* (ETA, 1978), and as a British Film Institute Educational Advisory Document (ed. L. Masterman, 1979).

1 TV and communication

'It is my theory that, while there are very few biased TV programmes, there is no such thing as an unbiased viewer', was the comment made to one of the authors by a well known broadcaster. His reaction to the audience and the public's reaction to media practitioners are startlingly similar–each vests in the other a form of irresponsibility providing a touching reminder of the communication gulf between them. The gulf is equally marked between practitioners and media researchers. Practitioners, engaged in the unremitting business of structuring events and opinions in intelligible form, criticise the language and limitations of much media research dogma, and their expectations of research are often too high. Researchers on the other hand commonly fail to understand the practical problems of media production. They too have high expectations of the objectivity that can be achieved in media communication and find, as the practitioners know already, that numerous uncontrolled biases obstruct the process.

Evidently TV is not necessarily a communication medium at all. Certainly no medium has the automatic ability to communicate of itself, though all have the potential to do so depending on their usage. When aural and visual images are transmitted via the electronic media they must be deliberately aimed and critically received or they will fall by the wayside. In view of the variety of interpretations placed upon the concept of communication, this is a tricky point to put across. For the makers of TV it is also a potentially dispiriting argument. It implies a chaos, a firing-off of messages with little or no knowledge of their reception. It also implies an audience polarised by a universe of interrelating differences, social and psychological. For practical purposes the viewpoint is clearly unhelpful, testifying to the complexity of the practitioners' task and the unlikelihood of their ever pleasing all of the people all of the time. Instead of facilitating true communication, therefore, in which the two sides of the process link together in at least a mutual sympathy, the media can too easily form a technological curtain through which the practitioners and their audience view one another with mutual suspicion. In considering this gloomy picture we shall attempt to define its bases and ways in which media research may help to remedy it.

Research and practice

How to develop a fruitful understanding between producers and public, researchers and producers is a major current problem. Since each camp, moreover, is internally polarised on questions of aims and methodology the prospect of a mutual sympathy seems at times remote. So let us repeat that TV is not necessarily a communication medium at all. McLuhan's vision of an harmonious global village, its members united by a common technological bond, has not as yet been realised. At best we may infer that TV has the potential for social and cultural influences of widely ranging types; and it is surely the hint of creative mastery underlying McLuhan's view of the media in this respect that has made it the only theoretical perspective in recent years to have penetrated the practitioners' natural defensiveness in theoretical matters (cf. Miller, 1971). As the populariser of numerous notions regarding the symbolic impact delivered by media, McLuhan directed practitioners' attention to the areas in which they are most vulnerable to critical attack. The challenge to researchers is to prove that they too can direct their attention to the same practical ends.

As the means to various practical outcomes, different approaches to media analysis have developed at different paces. Thus an approach via the classical information theory paradigm of Shannon and Weaver (1949) and Attneave (1959), while anticipating a possible future contribution at practical levels, has not yet moved beyond the level at which simple descriptive relationships between TV viewing tastes, age and education are measured (cf. Krull, Watt and Lichty, 1977). Baggaley and Duck (1976) indicate that the information theory style of measurement may yet serve a useful inferential role in media analysis; but it must avoid the assumptions that have dated it in other spheres of psychology and allow for the unpredictability of human reactions to information attributes. Its theoretical successor, signal detection theory (Swets, 1964), provides greater hope and a methodology that might even be used to extend the practical value of another recent arrival in media research, semiotics (cf. Baggaley and Duck, 1976, Chapter 6). In the latter connection an approach examining the creative factors underlying media communication has developed within schools of literary analysis and film criticism. Emphasising both the film producer's role, via the 'auteur' theory, and the interpretation of signs and meaning by the audience (Wollen, 1974), this perspective is rapidly increasing our awareness of the manipulative relationships linking them. The course of this development may be seen in relation to TV in a series of monographs published by the British Film Institute and discussed by Baggaley (1978).

In the initial contribution to the BFI series, Garnham (1973) considered the effects on the public of powerful organisational and cultural factors using TV as their agent. The link between medium and public that he describes is essentially one way, leaving the audience's capacity to react to such influences as a matter for separate examination. Dyer

(1973) then stressed the influence of light entertainment parameters upon TV production strategies. Again it is a one way process that he isolates rather than the two way mixture of factors observed in other communication situations: '. . . light entertainment intends a response, but response is not part of the situation at the moment of creation' (p.14). A similar determinist perspective is adopted in the study of TV's influence on a general election campaign by Pateman (1974).

In the fourth and fifth monographs, however (Buscombe et al., 1975; Collins, 1976), treating sports and news coverage, the viewers' response to TV begins to assume a greater significance. The manipulative powers of TV itself are still emphasised, but its subtle psychological bond with the audience becomes increasingly apparent: 'Instead of simply presenting reality to the viewer, whole and undigested, the visual media organise pictures into a kind of language which the viewer then "reads".' (Buscombe, 1975, p.3). From this position it is but a small step to considering the viewers' capacity for retaliation—cognitive if not physical—against media persuasiveness. Vaughan (1976) acknowledges the two way connection of TV and its viewers quite explicitly: 'What makes *The Ravenous Eye* (Shulman, 1975) so dispiriting to read is its neglect of the creative involvement of the viewer in construing a programme as meaningful, a creativity just as essential as that of the programme makers . . .' (p.10).

The two way viewpoint opens useful possibilities for the analysis of media effects alongside other communication phenomena. The fact that sociological and psychological precepts can complement one another in this analysis is an effective cornerstone in the construction of a unified 'communication studies' discipline. By broadening the theoretical frames of reference for media studies, so that TV for instance may be examined on a par with divers media—whether they be the hand that mediates through gesture or the pen that aids it with an italic flourish—one gains a richer research perspective, and one through which the theoretical view of electronic media can be tested at a new applied level.

Lines of social scientific enquiry have repeatedly followed a course from the theoretical to the applied in exactly this manner. Psychology, following an early period in which the hypothetical elements underlying human thought and behaviour were painstakingly segregated, was soon compelled to examine the conditions under which the whole operates and to adapt its theories accordingly. Educational theory, following a similar phase in which the supposed skills of teaching have been carefully classified, has considered also the psychological contexts provided for them by the learner, and the approaches required to deal with the differences between individual learners. Skinner (1938 et seq.) grasped the nettle in this respect and considered the ways in which individual behaviour could be 'shaped' for a more effective pay-off; and in the 1960s educational technologists began to explore in earnest the possibilities offered by communication media in this respect.

Within educational technology, ironically, the same process of trial and error development has now been repeated. At first it seemed that the annexing of modern technologies to the educational situation would reduce its problems at a stroke. Different media would provide different advantages in various situations, though once these were defined their correct application would be straightforward. Yet it has become obvious that the media do not invariably have an impact sufficient to overcome the basic learning problems, and no one medium—despite exhaustive research comparisons—has emerged as the most efficient even in quite narrow applications. In the matter of a medium's intrinsic impact the styles of usage are the independent variables (Baggaley, 1973) interacting with audience factors to determine the end effect. Immediately, we are back to a level intuitively obvious to the media producer. Even if we extend the analysis, disclosing the immense array of subjective and ideological obstacles to communication, he is not guaranteed to find a practical use for it. He has not been amused to recognise his foibles as depicted, for instance, by Glasgow Media Group's analysis of values and biases underlying TV news gathering (1976); while Altheide's equivalent study of American news broadcasting (1976), though rather more diplomatic in its approach, has also attracted primarily academic interest. Such studies demonstrate that TV journalists have feet of clay, but they are less forthcoming with hints of firmer footings to be taken.

Research indicating that a medium's impact is affected by the 'uses and gratifications' associated with it (Blumler and Katz, 1974) similarly exposes the audience as mortal. It indicates that the producer must become ever more ingenious if he is to deal with the vexatious viewer, and it charts numerous variables that may hinder him. Thus Lometti, Reeves and Bybee (1977) show that the impact of media material is conditioned by the needs of its audience for surveillance/entertainment, affective and behavioural guidance. The relative importance of dimensions in individual audience members is shown to vary with age. If research were also to offer ways to overcome such human problems, TV producers certainly appear anxious to test them in matters of production design. This being the level at which producers are practically equipped to act upon research-based advice, this is a critical level to which researchers should orientate.

Unfortunately, as Katz, Blumler and Gurevitch (1974) acknowledge, this particular approach has 'still barely advanced beyond a sort of charting and profiling activity' (p.25). It does not take, they add,

> the further step, which has hardly been ventured, of explanation. At issue here is the relationship between the unique 'grammar' of different media—that is, their specific technological and

aesthetic attributes–and the particular requirements of audience members that they are then capable, or incapable, of satisfying.

A stronger bond between the exponents of uses and gratifications research and the investigators of educational technology, busy in parallel, would have been of value to each. Both have independently studied the premise that human needs influence the satisfactions and benefits derived from the media, and have grappled with the problem of isolating specific needs from the human complex. In doing so they have reiterated a range of 'instinctive' forces charted between 1908 and 1924 by, for example, McDougall (12 instincts), Colvin and Bagley (25 instincts), Warren (26), Kirkpatrick (30) and Woodworth (110)! While able to show that different media functions are better at providing for certain human needs than others, both schools overlook the functional limitations of all typologies amply summarised by Bernard (1926):

> Several very diverse structures or behaviour patterns may have the same adjustment function, while the same behaviour patterns may at different times or in different situations perform antagonistic adjustment functions. (. . .) Neither is there a complete correlation between responses and stimuli, on the other hand . . . Consequently, these methods of classifying instincts are of but little value.

Since basic academic notions are condemned to cycle and recycle in perpetuity, uses and gratifications research has certainly served the beneficial role of alerting non-psychological areas of social science to the problems of individual diversity. Yet its proponents themselves admit that it must take a further step if it is to provide inferential evidence in the practical interest. It is therefore unfortunate that, in suggesting a further question to be asked in the process, they follow the early educational technologists in their assumption that individual media have absolute propensities only requiring to be tapped: 'Which, indeed, are the attributes that render some media more conducive than others to satisfying certain needs?' (Katz, Blumler and Gurevitch, 1974, p.25). Before the effective attributes of media have been ascertained and controlled there is no way we can know that some media are more appropriate for certain functions than others as this presupposes. Imaginative media control may produce silk purses from the most unpromising of materials. The value of research as an aid to practical media design will certainly be reduced as long as this assumption reigns.

For the present, however, the credence given to media comparisons in orthodox sociopsychological research still prompts numerous research questions, idealistically presupposing a tidy answer. Thus: what are the comparative effects of different communication media in, e.g. advertising and propagandist situations; or: how do we separate the influences of mediated and non-mediated communication? Since the exhaustion of such basic questions by the first wave of educational technology researchers they have been rephrased in a variety of more sophisticated forms, yielding equally inconclusive results (Clark, 1978). Thus: may television's efficiency as a communicator be increased for all learners via the use of certain instructional techniques (e.g. programmed learning)? Is the visual mode of communication more powerful than the verbal? Is there an interactive effect between presentation techniques and audience aptitudes? The latter development represents a completion of one question cycle within educational technology, as the researchers realise for themselves the Skinnerian principles that put them into business. It was discussed by Baggaley (1973) in an appeal for a more practical style of research based on questions concerning not only individual psychological differences (cf. Eysenck, 1953; Jamieson, 1973) but also the technical controls necessary to overcome them.

Eysenck has recently entered the media debate himself to point out that the perennial question of media and violence is insoluble unless personality differences of the extraversion/neuroticism type are taken into account (Eysenck and Nias, 1978). This observation will doubtless suggest new dimensions of media usage and gratification for study within sociology and social psychology; though it is important for us to realise that it has all been attempted before, in the literature of educational psychology as reviewed by Suchett-Kaye (1972). Once again the earlier efforts have shown that the parameter of psychological difference is but one of the determinants of media effects in practice; and since, in practice, all determinants are subject to constant change, typological approaches to any parameter are bound to have a limited predictive value. We must look not merely at the phenomena themselves but for the forces behind them, or they will probably move from sight as a star gradually passes out of the range of a too firmly fixed telescope. An inspection of the relationships between phenomena, conducted from various disciplinary reference points, may ultimately lead to a fuller understanding of the research and practical problems at hand than any one discipline can ever achieve on its own. We should search, accordingly, for interdisciplinary connections.

Images and meaning

In an account of communication by television, Baggaley and Duck (1976) have compared a number of separate viewpoints and their practical benefits. The variety of viewpoints adopted in media research, they indicate, has parallels in the broader theoretical study of communication phenomena. Attention to communication problems per se—as an explicitly practical matter—has been paid by engineering and social scientists since the war. The concepts of electrical engineering in particular, originally applied to human communication by Shannon and Weaver (1949), have played a formative role in the development of sociopsychological approaches to the subject and, as we have seen, they continue in currency within media research today.

Models of human communication based on engineering concepts alone, however, inadequately reflect its psychological complexity. Swept aside as 'noise', the anomalies and essential fallibility of human communication receive no illumination. In regular communication situations many different levels of the process are recognised, verbal and non-verbal factors interacting. While the linguist attempts to unravel the rules affecting verbal behaviour, the clinician, ethologist and anthropologist each refine a further set of rules at the non-verbal level. The explicit relationship between these levels has been explored by Bateson (1973). His seminal notion of the 'double-bind' indicates that information couched in both verbal and non-verbal terms combines for a collective effect upon those who receive them. If, as frequently, the signals conflict and require resolution, either the verbal or the non-verbal may take precedence.

When the impact of TV is considered in these terms a fresh set of hypotheses is indicated. The interaction between verbal and non-verbal effects of the medium has to be examined, and if necessary the skills of different disciplines pressed into service. It is insufficient to base analyses of TV's impact upon classifications of its verbal content alone, or to examine verbal and non-verbal messages in tandem without first making allowance for their capacities to exert independent effects. Lometti et al. (1977) point to the shortcomings of gratifications research in this respect:

> If subjects confound channel attributes with the content typically associated with the channel, then the usefulness of the information (yielded by the research) is significantly reduced. In the present study, as well as in the bulk of past research in the area, the researcher has no way of deciding which strategy was adopted by the respondents. . . . That is, channels and messages are likely confounded. (p.336)

Baggaley and Duck (1976) argue that the audience appreciation data collected by the major broadcasting organisations usually suffer from a similar problem, being analysed in terms of *ad hoc* programme groupings in which thematic, functional and technical characteristics of the content are confounded (cf. Dannheisser, 1975; Rowley, 1975). In restricted practical applications the classification of TV content according to function and theme may prove useful. Aspects of TV programme scheduling, for example, may be studied, as by Williams (1974), and the thematic balance in TV output within and between separate broadcasting organisations may be compared within the weekly cycle and over time. In teaching contexts this exercise can prove valuable, promoting an awareness of the strategies that schedulers use to involve the audience in media output (Masterman, 1978). In analysing output for such purposes at a week by week level the question of TV presentation technique, either within the cycle or within individual thematic categories, is actually of little significance as it plays no overt part in the balancing effects attempted by programme schedulers. But as the focus becomes more specific and day by day levels of analysis and comparison are attempted, technical factors become more prominent and their side effects less obvious. In analyses of TV's effects within specific contexts, therefore, the analytic approach to its content via programme function and theme alone must be broadened and the dynamic characteristics of (non-verbal) mediating techniques taken into account.

From different areas of the media field certain unifying concepts emerge for this purpose. The intentions behind communication, for example, have been studied with reference to media phenomena generally (cf. Jamieson, Thompson and Baggaley, 1976). The complementary processes of interpretation have attracted attention also, though with media reference this perspective has been too finely focused upon verbal content levels. More broadly directed at the non-verbal aspects of media presentation as well as the verbal, each of these two criteria pinpoints a major source of dynamic psychological variation for rationalisation henceforward.

From the viewpoints of descriptive and structural linguistics, stylistic analyses of the verbal and non-verbal rules of media expression have already been undertaken, notably by Metz (1974); and relationships between its narrative and formal elements have been indicated using a variety of methodologies. Semiotics, for example, the science of signs and symbols foreseen by de Saussure (1915), provides a theoretical framework capable of embracing both intention and interpretation viewpoints, also non-verbal phenomena as well as verbal (cf. Guiraud, 1975; Hawkes, 1977). Baggaley and Duck (1976) argue that the linguistic-based methods in media research at present suffer from the same theoretical and interpretative limitations as the approach to thematic content variables in isolation: they are based on a somewhat rigid taxonomic approach and on assumptions of a set of logical rules bearing no relationship to known psychological, social and cultural facts. Nonetheless, they are characterised by a rigorous intrinsic logic, and now require outward connections with social scientific lines of enquiry. With the addition of an empirical methodology, as available in the sciences of probability and signal detection, Baggaley and Duck suggest that the semiotic viewpoint for one can play a useful future role in the investigation of media effects.

Firstly, however, the possibility indicated earlier that TV exerts a range of independent non-verbal as well as verbal influences should be checked. In order to assess the likely nature and range of such influences, the same authors have considered the social and perceptual processes thought to underlie general communication phenomena. Following Kelly (1955) they suggest that man's basic need is for information and a structure by which to interpret it. Man is an enquirer, regulating his life and behaviour by forming hypotheses about his environment and the events within it, and by noting the criteria that most assist him. Individuals, moreover, may differ in this respect. When the information an individual receives is already familiar to him, a range of inner motivational and cognitive factors affects his reactions to it; and when it is not, as Festinger indicates (1954), the individual is forced to use more tenuous external criteria for evaluating the world around him, and a range of quite superficial cues can determine judgements not only of the personal interest value of incoming information but also of its credibility.

Cues to credibility, indicate Hovland and Weiss (1952), are regularly drawn from observations and expectations of the information's immediate source—i.e. the person or organisation conveying it (see also Hovland et al. 1953). Source credibility may persist long after the verbal content of the communication has been forgotten (Kelman and Hovland, 1953); and if both the subject matter and source of a communication are unknown, and 'maximally ambiguous' (Scheff, 1973), numerous non-verbal cues can take effect (cf. Argyle, 1969, 1975). When information is presented on TV, it is argued, these all important cues may actually be distorted, in two ways:

1 The TV situation frequently disturbs the behavioural cues conveyed in normal social interaction, causing the viewer to form distorted interpretations of the people, events and opinions mediated.
2 Since interpretations of interest value and credibility are known to be affected by the simultaneous and sequential contexts surrounding a stimulus (Kelley and Woodruff, 1956; Levy, 1960; Landy, 1972), the techniques conventionally used in mediation can exert a range of communication side effects in their own right.

The latter is a premise known to advertisers and propagandists, of course, whose depth and symbolist techniques aim to increase their persuasiveness in numerous ways (Packard, 1964). It is also well known in film criticism (Eisenstein, 1947; Pudovkin, 1958) and educational technology (Chu and Schramm, 1967; Schramm, 1971, 1975; Clark, 1975, 1978; Coldevin, 1976, 1979). In many modern TV contexts, however—the news and current affairs, for example—the possibility that presentation techniques affect the actual credibility of material is less well known, and production stylistics are commonly designed with a view to maintaining viewers' interest alone. Yet these are among the very contexts in which, featuring a parade of unfamiliar and ambiguous material daily, non-verbal effects on credibility may be predicted to occur. The techniques of TV presentation in such contexts, therefore, generally require close examination, since by their influence communication may, in theory at least, be jeopardised, with producers' intentions and viewers' interpretations diverging in practice.

The range of expressive techniques available to TV producers is vast and still expanding (see Baggaley and Duck, 1976, Chapters 2 and 4). Consequently the scope for unwitting context effects upon televised information is immense, particularly when the audience's critical awareness of its narrative meaning is low. A capacity for absorption in the imagery of the TV medium alone was inferred by Baggaley and Duck from the phenomenon of viewing 'flow'. (Highlighted by Williams (1974), flow describes the passive and unselective state in which viewers may be entertained by TV, quite independent of its intended range of meanings.) In general, they suggested, responses to material on TV are unlikely to operate on different psychological bases from those applying in normal perceptual situations, even though the medium itself may represent the material in a somewhat distorted manner. If viewers are aware of the distortions they may take them into account; otherwise the effects of TV subject matter and the manner in which it is conveyed are likely to be inseparable. In ambiguous situations the medium is thus a part determinant of the message, structuring it via the imagery it produces; another part determinant is naturally the viewer himself, who interprets the information structure according to his own personal criteria. In studying the presentation effects of TV, therefore, it is necessary to keep the viewer firmly in the picture. Attention to the psychology of the TV image by researchers in a range of conventional contexts would, it was argued, offer new possibilities for the design of effective production techniques in news, current affairs and educational TV, and for the instruction of viewers in the 'visual grammar' involved. The latter would afford viewers the opportunity to overcome the individual differences that obstruct their full use of televised information; it would also be required on ethical grounds, so that the public might learn to resist the less reputable propagandist influences of TV presentation.

To illustrate the non-verbal effects of media, Baggaley and Duck reported six experiments, in each of which an aspect of TV presentation was manipulated and overall effects upon viewers' attitudes measured (1976, Chapter 4). In the theoretical context that they served, these experiments were by no means central to the work, though they indicated a methodology for possible future application. The present book aims to develop and extend this methodology in a variety of contexts. In the course of the research an attempt has been made to respect the essential dynamism of TV's impact, and to record variations due not only to its practical and artistic conventions but also to inconsistencies within the audience's reactions to them. As communication technology develops, the dynamics of TV will continue to fluctuate. They may be harnessed for the public good as long as the effects of presentation technique are predictable by the producer and viewer alike. As the following section argues, the description, prediction and eventual control of such processes must be based upon a correspondingly flexible research approach.

Investigating visual impact

In the development of an effective communication medium it is evident that numerous independent approaches impinge at once. Organisational, linguistic, cultural, ideological viewpoints all fulfil research roles at different levels. No one approach surpasses the others in being able to deal with all of the problems tackled by media research, though approaches on the whole do differ in terms of their ready practical applicability. Moreover, much research has been needlessly devoted to questions already studied in other academic fields, with few links established between disciplines guiding the development of media practices in the dynamic human context.

Since publication of the line taken by Baggaley and Duck (1976) certain of its elements have been spontaneously echoed elsewhere. The educational use of media to overcome the individual psychological differences obstructing learning efficiency has been labelled 'compensatory' (Clark, 1978). In devising techniques to compensate rather than to cater for learning problems new approaches to more effective group instruction are indicated in the manner discussed earlier. Accordingly, current research (e.g. Salomon and Cohen, 1977) predicts that the traditional heavy emphasis within educational technology upon individual and self-instructive media procedures will be relieved in future. Prospects are thus good for the design of effective distance learning presentations catering for large sections of the populace in a uniform manner.

Within the literary tradition, Fiske and Hartley (1978) have offered a 'bardic' notion of TV. Their view that TV performs a set of cultural functions identical to those of the classical bard lays useful emphases on the oral techniques used to structure mediated events for a particular interpretation, and on the dynamic context in which the media practitioner operates. The 'dynamic' perspective, as all others emphasising the manipulative capacities of mediating technique, is naturally open to suspicion from the practitioners themselves. Their experience of research into mediation has taught them to expect 'the conspiracy theory', the implication that their biases and subjective values are consciously applied. In the present research, however, this is specifically disavowed: the biases tapped are as likely to stem from viewers' prejudices as from any belonging to the producers, and they are generally assumed to be unwitting (cf. Baggaley and Duck, 1976, p.117).

Exclusions from the latters' viewpoint may be criticised—as by Sherrington (1978), who indicates their scant reference to cross-cultural studies of media symbolism as in India and Africa; and by Corner (1978) who stresses the important effects of the broadcasting system in the communication process. We accept the importance of these areas in media theory generally, but stress that inclusion of material in this and the previous volume has been primarily governed by considerations of its ready applicability to problems of production design. Presentation research conducted on a similar basis is now under way in Canadian contexts (Coldevin, 1978; Sullivan et al., 1978-79), and comparison of these findings with those obtained in the British TV culture will be interesting.

The importance of the broadcasting system in media analysis, indicated by Corner, has also been raised in the TV violence debate by Smith (1978). Effects cannot be attributed to the 'nature of the TV medium' per se, he suggests, but to the organisational pressures upon its producers: 'A broadcasting system that is inadequately financed, or badly governed, or made to compete feverishly internally, will be tempted into extraneous violence and pornography, because these can be obtained cheaply and effortlessly' (p.14). As we have stressed, a quite independent set of factors may indeed be predicted to stem from the formal 'nature of the medium' also, and the series of experiments reported below was conducted in order to test the practical significance of these factors. The experiments are based on the following empirical criteria.

1 *Theoretical context.* Our main concern in the empirical work to date has been to elucidate the basic hypothesis, discussed above, that the stylistic techniques used in TV presentation can affect viewers' reactions to the credibility of the persons presented as well as to their interest value. (Experiments 1 to 24 explore this notion specifically.) The broader theoretical context, however, concedes that particular reactions to such qualities may be highly idiosyncratic, according to the characteristic personal constructs of the individual. The latter hypothesis, couched in terms of personal construct theory (Kelly, 1955) has already been studied in detail in non-media contexts by, for instance, Duck (1973); and Baggaley and Duck (1976) have indicated the value of Kelly's 'repertory grid' technique as an empirical means of testing the theory in relation to TV viewing. The reptest has particular power in the detailed investigation of an individual subject's constructs (see Slater, 1969), making no assumptions as to those he may employ in different situations. Nor, consistent with Kellian

principles, does it assume that the members of a subject group share any of each other's constructs: '... no two adult individuals can ever be the same in cognitive structure. Nevertheless ... some overlap is both necessary and certain in order to avoid complete isolation and solipsism' (Duck, 1973, p.24). In the majority of our experiments we take advantage of this fact, concentrating for convenience upon majority reactions to the TV image; but in Chapter 6 we describe a technique for defining the subsections of a population or subject 'mass'–i.e. its degree of differentiation or cohesiveness. And in the sections of that chapter describing Experiments 25 to 29 we demonstrate applications of this technique in the analysis of effects stemming from the general nature of the TV image and from the factors endemic to its audience simultaneously.

2 *Experimental hypotheses.* A wide range of presentation options taken in common TV practice is examined–camera angle, visual background detail and editing technique, dubbed music, subtitles, autocue, styles of performance, etc.–and their effects are observed on contrasting groups of adults and children. Although the experimental literature concerning attitude formation suggests that judgements of credibility and interest value may be particularly susceptible to presentation influence, especially in ambiguous situations, the direction of effects and the possibility of interactions between them are at this stage largely unpredictable. Until a sufficient number of the independent variables within a research field have been exposed empirically the formation of clear cut *a priori* hypotheses is problematic. Each experiment that follows, though prompted by intuitive suspicions regarding the influential nature of certain variables, is therefore based on a 'null' hypothesis: i.e. that the variable(s) in question will have no effect at all. An appropriately stringent statistical criterion for the rejection of this hypothesis is applied in the analysis of results (see p.20).

3 *Design and TV materials.* In each experiment televised material embodying, where appropriate, a variation or variations of presentation style is presented to subjects for their reactions on certain bases. The material has either been produced using the closed circuit TV facilities of Liverpool University or taken from national network sources. In the experiments where effects of a particular stylistic variable or variables are studied the material is designed so that only the factor(s) under scrutiny are varied, and all others, e.g. verbal content and performance, are kept constant. Using electronic and editing resources separate copies of the material are prepared on videotape, identical in all but the experimental respect: they are

then shown to separate subject populations. For a check on the significance of certain variables across a contrasting and unpredictable range of situations, as in 'real life', they are tested separately, in combination, and as indicated above on different types of subject.

The major design criterion governing all experiments has been that no subject should be able to recognise the experimental manipulation, or to structure his/her responses according to a knowledge of experimental intentions. Thus no subject in an experiment ever sees more than one of the conditions for comparison; and, since the general hypothesis stresses the particular powers of presentation in ambiguous situations, the subject matter and performers in the material are generally selected so as to be unfamiliar. In the experimental 'blind' situation thus created such artefactual variables as may be introduced by experimental design and administration procedures (see p. 17) are, if not controlled, common to all conditions. When, despite these possible contaminations, a recognisable main effect is yielded by the experimental manipulations nonetheless, we may conclude that it deserves to find, in Kaiser and Caffrey's terms (1965), 'its place in the sun'.

The materials, when not obtained by off-air recording, have been prepared using Link 103 cameras and Ampex 5803 videotape recorders, on 1 inch high density videotape (3M) for copying on to ½ inch Philips video-cassettes. In Experiments 12, 13, 16 and 24 to 29, professional TV presenters were judged; in the remainder the performers were teaching staff at Liverpool University and/or actors in the Merseyside amateur theatre. Only in Experiments 28 and 29 was the performer known to the general public. The types of verbal material used in the experiments are indicated in Appendices 1 and 2, obtainable from the author.

4 *Experimental measures.* In his attempt to measure the effects of any stimulus the researcher has several alternatives. He can ask subjects to describe their reactions freely, which in the present 'blind' situation concerning influences presumed to be unconscious would clearly be inappropriate. He may use direct questionnaire techniques, though the risk of involving his own subjective predilections is high using this approach: direct questioning runs the risk of 'cuing' the subject as to the experiment's purpose and appropriate reactions, and where possible should only be used as a final measure for comparison with other methods used independently (Baggaley, 1973). He can apply measures of, for example, immediate and delayed recall which usually demand the rather elusive guarantee that subjects are totally unfamiliar with the type of information

presented, or alternatively involve comparisons of pre-test and post-test performance, which in turn involve 'cuing' effects (Hartley, Holt and Swain, 1970). Or he can attempt to predict the subjects' future action tendencies via, for instance, attitude measurements.

The limitations of attitude measurement are similar to those of general questionnaire methods. If the scales of measurement are dictated by the experimenter, responses can be suggested to the subject that he would not make of his own accord. If on the other hand an open ended technique is used allowing the subject to create his own attitude scales, statistical comparisons within a group of subjects become difficult. (Kelly's repertory grid technique offers useful possibilities in the latter connection though is relatively cumbersome in its administration.)

However, it is a relatively simple matter when using attitude measures for the experimenter to disguise any expectations he may have of particular effects by including in the test measures that he presumes to be irrelevant. In this way he is also able to detect whether an experimental variable affects subjects' reactions indiscriminately or quite specifically. Moreover, as indicated by Kelman and Hovland (1953), the predictive value of attitude measures may persist into the long term specifically in relation to perceptions of source credibility. The measures may thus be used in attempts to tap the fundamental 'sleeper' effects of communication—the unconscious effects upon subjects that do not become apparent until unspecified subsequent occasions when spontaneously aroused (Aronson, 1973). In the present research we have decided to emphasise the effects of TV presentation upon viewers' attitudinal measures accordingly.

The technique most widely used in attitude measurement is the 'semantic differential' originated by Osgood, Suci and Tannenbaum (1957): it is easily administered and analysed, and permits the testing of a wide range of hypotheses (Baggaley and Duck, 1976, p.85). The degrees of meaning that a subject vests in a stimulus are measured on a series of bipolar adjectival scales commonly containing either five or seven points. Attitudes on a range of bases (e.g. FRIENDLY/HOSTILE, FAIR/UNFAIR are recorded between the extremes of the scale, in which the midpoint represents an undecided or neutral response (see also p.19). Either extreme may express the positive or negative pole of a particular attitude construct as required. In each of the present experiments the subjects' reactions to a TV performer (the immediate 'information source') were invited in this manner on a set of seven point scales presented to them after the videotape presentation. The diversity of uses of the semantic differential technique in the social sciences, including communication research, is indicated by Warr and Knapper (1968) and by Snider and Osgood (1969).

The pioneering investigations of meaning measurement by Osgood and his colleagues (1957) indicated that meaning is generally attributed according to three major factors:

(a) The *evaluative,* comprising judgements of basic 'goodness' and 'badness' in the object, person or event perceived.

(b) The *potency* factor, concerning the qualities associated with power and strength in particular contexts.

(c) The *activity* factor, concerned with qualities such as quickness, excitement and warmth.

As Osgood et al. themselves point out, the meanings behind particular judgements may vary from context to context, and in restricted applications 'the nature, order and magnitude of the factors may change': 'For example, when judgements are limited to socio-political concepts . . . there seems to be a coalescence of the second and third factors into what might be called a "dynamism" factor' (p.74). The need to use the semantic differential flexibly and with due regard for the situation at hand is thus clearly indicated.

In the present context, therefore, different combinations of seven point scales considered to represent each of the major classical dimensions of meaning are applied, with the inclusion of certain other scales (e.g. BELIEVING/SCEPTICAL, RATIONAL/INTUITIVE found by Osgood et al. to have less firmly prescribed meanings: a useful thesaurus of scales is given in their text). The scales used throughout the present research are indicated in Appendix 3 obtainable from the author. Scales that prove yielding in one connection have been subsequently administered in others and evidence thereby collected regarding the consistency of particular effects overall.

In the presentation of scales to subjects via the data collection sheet the opposing 'positive' and 'negative' poles of each have been determined intuitively, and the 'positive/negative' and 'negative/positive' orders of scales on the sheet varied randomly so as to minimise chances of artefactual response bias (Cronbach, 1956). The instructions to subjects in the rating tasks are given on p.19). In Chapter 4 the use of other behavioural measures is reported.

5 *Subjects.* Subjects in the majority of experiments were undergraduate students at Liverpool and Lancaster Universities—eighteen to twenty-two years of age. Many of them being social science students who may be assumed to have a certain insight into experimental motives, the need for a 'blind' situation in which each subject rates only one of the experimental conditions to be compared is crucial. No subject has been told of the experimental manipulation or its purpose unless all data for the experiment have been collected. Since the reactions of students in general cannot be assumed without test to be representative of other subject types, a variety of checks using children, secretarial and clerical staff has been made. All subjects participated in the experiments anonymously. (Regarding the decision to concentrate primarily on student data see p.49).

In each experiment data are collected from a sufficient number of subjects per condition to permit the analysis of results according to standard criteria for statistical significance. Subject numbers range between 10 and 28 per condition in different experiments: in each condition equal or approximately equal numbers of male and female subjects are used. It is naturally in the experimenter's interest to use as large a number of subjects in each sample as possible, since the statistical criteria applied in dealing with low numbers are correspondingly more rigorous and the chances of observing a significant effect are reduced. However, as Bruning and Kintz (1968, Chapter 2) point out, most experiments based on an analysis of variance model as here (see pp.19-20) use ten or more subjects per condition as we have done.

6 *Administration.* Consistent with the design principles discussed earlier, the experiments have been administered in an informal manner bearing as little resemblance to the 'clinical laboratory' style of experimentation as possible. In the 100 years since the foundation of the first psychological laboratory at Leipzig by Wilhelm Wundt, psychologists have become familiar with popular if insubstantial assumptions regarding their experimental methods and if sensible anticipate them. A typical general criticism is made, for instance, by Smith (1978), who in disparaging the conclusions of Eysenck and Nias (1978) regarding the media's relationship to social violence suggests that: '(W)hile the subject sits before the white-coated figure of Dr. Eysenck, media violence may appear to increase his aggression and sexual libido; but in the context of the whole of his television viewing . . . quite different principles are at work' (p.14). In view of our own certainty, discussed earlier, that numerous extraneous factors are indeed at work quite unpredictably in the viewing

experience, the bold step has been taken in the administration of our experiments of making no attempt to control them! Subjects have been tested in one's, two's and small groups until ten or more of them have been subjected to each condition. The results of different test sessions are pooled to form the data of individual conditions accordingly; small rooms, larger rooms and corridors have been used indiscriminately; the experimenter has sometimes remained in the room otherwise occupied and at other times walked away; other people have been allowed to pass by, to talk or play music in adjoining rooms according to whim; in short no steps to falsify or control the viewing environment have been made, though administrative rulings that might affect the conduct of one experimental condition and not others have been carefully avoided. It has been assumed that worthwhile comparative effects between conditions are strong enough to emerge despite the barrage of contaminating effects—halo, Hawthorne, social contagion, etc. (cf. Cronbach, 1956)—that the real world produces. All such contaminations may be presumed in any case to influence the contrasting conditions in an experiment equally, as indicated on p.14. The experimental situation need not be an obstacle to applicable psychological research as commonly feared, any more than in that of the physicist, biologist or chemist.

In the present research little purpose could be served by creating an artificial 'laboratory' situation anyway. Eysenck (1978), replying to Smith, indicates 'that laboratory experiments are done to test specific hypotheses, and that the more specific these hypotheses, the more readily they are susceptible to laboratory testing'. Since we have assiduously avoided forming specific *a priori* hypotheses in this research (see p.13) any steps we might take to screen out particular influences upon our data would be as likely to go against our interests as in favour of them. The criterion by which the value of any result then remains to be judged is the extent to which it serves the theoretical context of the research (see p.12) and a useful purpose.

Having been contacted in a variety of ways, therefore—via notices inviting them to help research by watching TV, in captive groups at school, during lectures and lunch hours, in the street, etc.—subjects are told that they will not be required for more than a few minutes and that they should simply relax and watch a brief TV extract. After seeing the extract (between one and five minutes in length depending on the experiment) each subject is handed an A4 typed set of attitude scales labelled 'Impression Sheet', with the bidding that their first impressions only are required and that undue effort in the task is therefore unnecessary.

The impression sheet begins with the following instructions: 'Please indicate your impression of the' . . . . . (lecturer/interviewer/speaker/performer/etc.) . . . 'who appeared on the videotape you have just seen. Use the following scales, circling the number on each which most accurately reflects your opinion.

For example, if you thought the person was "quite kind", you would circle number 2 on the scale below:

KIND	1	(2)	3	4	5	6	7	CRUEL
		Quite kind	Moderately kind	Can't decide what he was	Moderately cruel	Quite cruel		

Note: The scales that follow are not written out in full, but the numbers in each case represent the same graded steps as above.'

(The individual scales then follow in random order: see Appendix 3 obtainable from the author.)

At the foot of the impression sheet is the question: 'Have you ever seen the person before? YES/NO'. In view of the general hypothesis (above) regarding the role of source familiarity in this research, the data of any subject believing this possible have been excluded from the analysis. The subjects who, on rare occasions, have failed to respond on more than one half of the scales, are also excluded. However, if a subject fails to respond on one or two of the scales only, a 'missing value' of 4 (the scalar midpoint) is inserted retrospectively.

Following the basic attitude measurement certain other measures have occasionally been administered, as in Experiments 6 to 8. Finally, a general all purpose explanation about viewers' reactions to TV is given to the subjects seeming to want it. They are then thanked and allowed to leave.

7 *Analysis of results.* In the analyses of attitude ratings the scores on all scales which had been presented to subjects in a 'negative/positive' sequence are reversed, so that the most extreme 'positive' rating on all scales is a 1 and that at the opposite pole a 7. (When results are tabulated, the positive pole alone is quoted for brevity; full detail is available in Appendix 3.) The scores in the various conditions per experiment are then usually compared via standard statistical tests for the significance of differences, in the manner indicated by Baggaley and Duck (1976). Thus, in the basic situation in which subjects' scores in two conditions are compared, a Student's t-test for the significance of differences between independent means is used (Bruning and Kintz, 1968, Chapter 1). Via this and the other tests for statistical difference (e.g. analyses of variance) used throughout these experiments, the arithmetic mean of subjects'

scores in an individual condition and the variation due to uncontrolled differences between individual subjects within the condition are both taken into account—and not, as a reviewer of the preceding volume has imagined, the means alone! The means of scores found significantly different on this basis and the levels of statistical significance themselves are tabulated, as in the previous book, in the context of each experimental report; statistical 'degrees of freedom' (Garrett, 1964, pp.194-5) are also given.

In the analyses comparing more than two conditions simultaneously, the following styles of variance analysis are used (cf. Winer, 1971, p.445):

(a) when the subject numbers in different groups are equal, a balanced design;

(b) when they are unequal, an unweighted means design;

(c) when the experimental conditions are classified on one dimension only (e.g. the stimulus material is varied in only one basic respect such as visual background detail though three types or levels of background are employed), a one way design;

(d) when the conditions can be classified on two dimensions (e.g. visual background detail and edited detail simultaneously), a two way design.

If the variance analysis indicates the existence of a significant difference or differences between unspecified experimental conditions, further t-tests are conducted as multiple comparison analyses aiming to determine the precise conditions involved. In each of the latter analyses the stringent two tailed significance criterion is applied (Edwards, 1965, pp.95-7), as essential in the open ended empirical situation in which no directional prediction is made (see p.13).

Other tests used in particular experiments (chi-square, factor analysis, etc.) are reported in the appropriate chapters: regarding the basic rationale for factor analysis, see Experiment 19. All statistics, excepting the relative weighting quotient of Chapter 4 and the magnitude correlation coefficient of Chapter 6, are compatible with the procedures of the standard SPSS computing package (Nie, Hull et al., 1975).

8 *Conclusions.* In our analysis and discussion of each set of results below we have more commonly concentrated on drawing conclusions of an applied rather than theoretical type, though on occasion—for example Experiments 11 and 14—we have been tempted to pursue

a slightly more rarified research question. We have aimed to cover a wide variety of different possibilities concerning image effects rather than to fix upon one or two variables alone in searching theoretical detail. As one set of results is examined it may suggest a tangent more worthy of exploration than the original question itself; and in this situation we have commonly continued by exploring the new tangent even though questions may remain unanswered regarding the old one. Therefore, although we primarily aim to provide practical answers in our discussion sections, we also try to raise questions for further study. We have been governed in this attempt by the pragmatic criteria emphasised in the earlier sections of this chapter and we have consequently been quite cautious in the conclusions based on isolated experiments, a requisite in this field stressed by Baggaley and Duck (1976):

> The experiments do not, therefore, claim to be exhaustive, nor do we suppose our results to indicate the relative efficiency of particular production techniques in general terms. The effects we report are *unique to the particular contexts from which they are derived* and the use of these techniques in other production contexts henceforward may yield entirely different results. (p.83)

In our final conclusions, however (Chapter 7), we do invite the reader to notice highly consistent effects now emerging across the range of experiments. These suggest that the dynamism of TV is actually more amenable to prediction when the two way effects of TV production and viewing are considered interactively than Baggaley and Duck have previously supposed.

2 Levels of visual awareness

Two basic types of TV presentation effect have been supposed and now need to be tested empirically. On the one hand it is argued that TV frequently disturbs the cues conveyed about people, events and opinions in normal social interaction, causing the viewer to make distorted interpretations of them. On the other hand it is possible that the visual techniques used to present information via TV provide a further range of cues in their own right, affecting viewers' judgements of, for instance, interest value and source credibility. The first six experiments reported in this chapter exemplify each of these premises and pose questions for subsequent investigation. As the results of this early experimental series have been published elsewhere (Baggaley and Duck, 1976, 1979) we do not retabulate the statistical findings here unless, of course, extending the previous work as in Experiments 6 to 8. In general we raise several implications of these experiments that the previous book has not covered. The work from Experiment 7 onwards is reported here for the first time.

This preliminary set of studies indicates various levels of audience sensitivity to the TV image, a number of the presentation cues responded to, and steps that may be taken to harness image effects in, for example, educational broadcasting.

Experiment 1: Use of notes

Our first impressions of other individuals in face to face interaction are formed on notoriously insubstantial grounds (Warr and Knapper, 1968). Assumptions of a person's authority and reliability are based on the evidence of his clothing, length of hair, bearing and mannerisms, and on other very marginal cues. When a person is viewed in the unnatural conditions imposed on him by the highly selective medium of TV, many of these usual clues to his character are concealed and, in their absence, viewers' sensitivity to the suggestions about a performer arising from the visual context in which he is presented is likely to be enhanced. The first experiment checks the possibility that viewers' attitudes to a TV performer may indeed be influenced by the variation of a simple presentation cue in this manner. The experimental manipulation, as in all of the experiments to follow, is of the type dictated by the members of a TV production team in the normal course of their work. It is effected by a simple change of camera angle.

Procedure

A three minute lecture type item was televised on the subject of Edgar Alan Poe. Two recordings of the lecture were made simultaneously, using two cameras placed side by side. The recordings were identical except for the slight variation in camera angle seen in Figures 2.1 and 2.2. (For further detail see Baggaley and Duck, 1976, pp.86-8.)

Figure 2.1 Use of concealed notes by a TV lecturer – Experiment 1

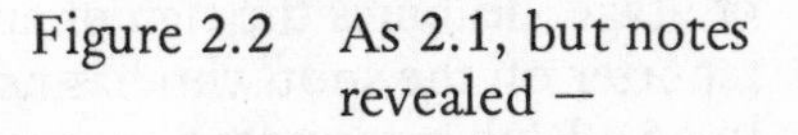

Figure 2.2 As 2.1, but notes revealed –

While one camera framed the speaker centrally, the other framed him more to the left of the picture bringing into vision the sheet of notes on his knee. The two recordings were each shown to separate student audiences, 11 subjects in each (N = 11, 11), and their assessments of the speaker were obtained on 14 attitude scales by means of the semantic differential technique (see Chapter 1).

Results

The two groups' scores on each scale were compared statistically (t-test for the significance of a difference between two independent means). Viewers of the tape revealing the speaker's access to notes found him significantly less FAIR and more CONFUSING than the viewers who did not see the notes. Since both audiences in this experiment saw and heard precisely the same performance, the difference in their assessments of the performer's fairness and straightforwardness can only derive from this most superficial of cues about him.

'Significance' in this situation is demonstrated according to criteria indicating the probability (P) that differences between the groups' scores have occurred by chance; the differences reported above were judged likely to occur by chance less than once in 100 occasions ($P < 0.01$). If results show that a difference might occur by chance more than five times in 100, it is deemed non-significant.

Discussion

Evidence of a TV lecturer's use of notes diminished viewers' perceptions of his fairness and straightforwardness.

When viewers were questioned about the performance it became evident that the minor detail as to notes had indeed been instrumental in this process, though at a quite unconscious level. The audience to whom the notes were visible assumed that when not addressing the camera the speaker was referring to them. Viewers who had not seen the notes interpreted the deviation of his gaze from the camera as due to pensive deliberation. Each of these assumptions was a tacit, automatic reading of the performer's behaviour, and neither group of viewers was aware of its influence on their attitudes towards him. Yet on these assumptions the further inferences regarding his fairness and straightforwardness as a lecturer were evidently based; and it is probable that attitudes to the content of his lecture will have been affected in turn—a possibility considered in relation to other variables in Chapter 4.

Fiske and Hartley (1978) discuss this result in terms of 'visual semantics'. When verbal meaning is analysed according to linguistic rules the smallest units by which meaning may be conveyed are known as 'semes' (Greimas, 1966). Discussing the 'lecture notes' detail as a seme capable of association with degrees of fairness and straightforwardness, Fiske and Hartley indicate that it may gain a different meaning on TV from its meaning in non-TV contexts: 'It may well be that the expert speaker on television uses an autocue, and thus "lecture notes", by convention or usage, becomes the sign of an amateur . . . or they may signify the reporter on the spot who has not had time to write a "proper" report' (p.65). Both interpretations are consistent with the assumption by Baggaley and Duck (1976) that such effects are determined by conventions of presentation style and may be context specific. The first explanation, however, seems the more feasible. It allows that viewers who are not aware of teleprompting or 'autocue' technology may assume the speaker without notes to be improvising, a suitably expert qualification. It is less likely that viewers will hold use of notes against a TV reporter, in view of our results in Experiments 2 and 13.

Experiment 2: Adding a background

In the straightforward TV report situation where, for lack of more appropriate visual information, the lecturer or studio presenter features in vision as well as sound, the producer has the choice of setting him against either a blank or a decorative background. He will often opt for the decorative background in order to increase visual interest. The modern electronic process of 'keying' permits the producer to combine the image of the presenter with the image from a second camera looking at, for example, a photograph or slide. The illusion is created that the presenter is set against a particular background, while problems of studio set construction are avoided.

Yet production staff should guard against adverse effects—such as the distracting influence of an irrelevant or abstract background—which in educational TV particularly could undermine the presentation's impact. While the latter is a simple effect to predict—and one of which any producer would be aware—the scope for misuse of such techniques is immense.

Previous research into visual background effects (Ellery, 1959; Schlater, 1970; Barrington, 1972) has been inconclusive; it has measured presentation effectiveness in the relatively fickle and problematic terms of information retention alone (see Chapter 1) overlooking, as Coldevin (1976) points out, other aspects of the viewing experience possibly capable of more reliable measurement. The present experiment, therefore, used the semantic differential methodology in seeking to determine whether the insertion of a background image has the potential for more subtle effects than might commonly be supposed.

Procedure

A news type item reporting an archaeological dig by members of Liverpool University was scripted by a professional TV newswriter, and two televised versions of the item were prepared. The two recordings were made simultaneously using two cameras immediately adjacent and separate recording circuits. The image (of a 'head and shoulders' close up on the presenter) remained constant throughout the recording; the presenter read his script from an autocue device between the two camera lenses (see Experiment 13) and appeared to address each of them directly. The two camera angles were indistinguishable, and the two recorded versions differed perceptibly in only one respect: that in one the presenter was seen against a plain grey background (Figure 2.3) while in the other he appeared to be in front of a large scale studio photograph of the archaeological site in question, effectively inserted by the keying process despite the problems of keying in monochrome (Figure 2.4). The recorded item lasted seventy seconds.

Figure 2.3 TV newsreader: plain background – Experiment 2

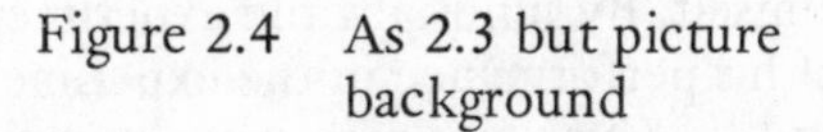

Figure 2.4 As 2.3 but picture background

The two versions of the item were presented to separate student groups (N = 13, 12) who were then asked to indicate their impressions of the televised presenter on 13 adjectival scales as in Experiment 1. The scales related primarily to qualities of personal attractiveness and expertise.

Results

The significant differences in response between the two student groups yielded by t-tests on each scale as above indicate that when the presenter appeared against the keyed background he was judged to be more HONEST, PROFOUND, RELIABLE and FAIR ($P < 0.05$ in each) than when seen against the plain background. On the other hand he was not regarded as significantly more INFORMED or EXPERT, and the overall ratings of his interest value on the two tapes were virtually identical.

The insertion of the picture background seems to have had less influence in intuitively obvious respects, therefore, and more upon the performer's level of personal credibility—a normally quite unpredicted effect upon the amount of trust which the two groups were prepared to invest in him.

Discussion

The 'keyed' insertion of a picture background increased a newsreader's perceived credibility rather than his simple interest value.

Coldevin (1978a) suggests that the absence of an effect upon interest value in this experiment may have been due to the fact that only one visual background was used during the seventy second period studied. In a replication featuring highlights of the 1976 Montreal Olympic Games, he compared reactions to several picture conditions and a plain one as above. In the picture conditions, however, he varied the visuals used three times within ninety seconds: in a full picture setting both the impact of the information and of its presenter were rated consistently more favourably than in a plain setting. The verbal information itself in Coldevin's study was rated as significantly more COMPLETE in the picture background situation, and more IMPORTANT. A further condition—in which a corner inserted picture was used as per popular TV news reporting convention (intermediate as it were to the plain and full picture conditions)—yielded correspondingly moderate effects enhancing the speaker's perceived honesty and clarity.

The combined results of the Baggaley and Duck (1976) and Coldevin (1978) studies confirm the powerful effects that visual context may exert upon viewers' attitudes, and the care to be taken in their use. Of course, the precise direction of such effects may be as variable as the nature of the background detail used (cf. Barrington, 1972; Canter, West and Wools, 1974); and in Experiment 12 we explore this possibility further. The particular results of Baggaley and Duck may also be explained in relation to the visual appearance and vocal manner of the performer himself. By hindsight the experimenters felt that these separate aspects of his performance in the experimental recording may have conflicted; and that his voice may have been the main determinant of favourable reactions to him while his visual appearance was less favourable—probably in view of his steely reliance on the autocue device which he had used in maintaining a smooth verbal delivery. Whether or not a cuing device is available in the studio the additional value of a desk script is thus indicated. For unless the speaker may be assumed by his audience to be an expert in the actual information he conveys—which the conventional TV reporter is not—his need to refer to notes to some extent will be expected in any case. The association of 'notes' and lack of preparation, indicated by Fiske and Hartley (1978) in relation to Experiment 1, is thus less likely to underlie reactions to the TV reporter than it is to the subject specialist.

While further data are needed to confirm this particular interpretation (cf. Experiments 11 and 13) ample previous findings regarding the separate processing capacities of eye and ear (e.g. by Broadbent, 1958; Travers, 1964, 1966; Coldevin, 1975a) certainly indicate that the auditory and visual content of a communication must be carefully matched in order to prevent the receiver's attention from shifting between them and favouring one at the expense of the other. If decorative rather than informative, the insertion of keyed visual detail may serve to distract an audience's attention from the speaker's visual presentation, leaving his auditory channel as the main communicator. Of course, when an educational producer is compelled to use an expert presenter who has difficulty in projecting a favourable visual image, this technique may be used to good effect. For by an effect on the presentation's general arousal potential the presence of background detail in such a situation may improve its value even when irrelevant to the subject under discussion.

In an educational context where several media are available the natural response to a situation such as this should, of course, be to question whether TV is necessarily the most appropriate. In many situations a televisual presentation conveys nothing that the audiotape and/or visual slide projection media, if carefully used, do not convey more economically. However, in presentations where a televisual treatment is essential on, for instance, logistic grounds though less so in other respects, the decorative uses of visual keying may serve a valuable function despite the fact that the auditory channel carries the actual information for learning. The main question of practical concern in such situations becomes less one of comparisons between media than of the techniques for using each one to best advantage; and to TV producers requiring, quite simply, information as to how to improve their uses of the single medium at their disposal, questions of media comparison (still popular with academic researchers) are indeed purely academic.

Experiment 3: Front view or profile?

From the first two experiments it is apparent that numerous interrelating production variables may affect viewers' attitudes to a speaker and his information at once. Further variables such as speed of delivery, posture and gesture may each have similarly interacting effects. Yet in each of the previous experiments all factors such as these were controlled, while by careful camerawork variations were made in aspects of the visual imagery alone. Variations in peripheral and background detail each affected a speaker's apparent credibility to a significant degree; and it seems clear that the formal nature of TV can exert its own, quite independent influences on the perceptions of those it represents. The third experiment pursues this hypothesis, and the extent to which audience reactions may be affected by the very angle at which a performer is presented, as in Experiment 1.

Procedure

A one minute address was televised on the subject of law and order. Two simultaneous recordings were made presenting a head and shoulders view of the speaker from two angles (Figure 2.5). Seated against a curtain background he addressed one camera, with autocue, directly (Figure 2.6) and was shown by the other in half profile (Figure 2.7) as though participating in a discussion. Focal length in the two recordings was identical, and in each the shot remained unchanged throughout the extract. (Further detail is given by Baggaley and Duck, 1976, pp.90-3.)

Results

Audience reactions to the two presentations (N = 10, 10) were obtained on 14 rating scales as previously; and following t-tests assessments of the speaker were indeed seen to be affected by the angle (direct vs. profile) at which he was presented. In the half profile condition significantly more favourable ratings on the RELIABLE and EXPERT scales were obtained ($P < 0.05$ in each); and on 12 of the 14 scales generally the half profile condition drew more favourable ratings than the direct condition, even though the differences between them were not always statistically significant. (NB: The subject population used here was a slightly augmented version of that reported by Baggaley and Duck previously, though the results remain the same.)

Discussion

When seen to address the camera directly a performer was considered less reliable and expert than when seen in profile.

In popular usage the effects of camera angle on a performer's appearance have long been exploited—a low angled shot, for example, making a small man seem larger and a big man bigger. Producers well know that the qualities of a man's physical appearance suggest further characteristics—of, for example, power, activity or otherwise—which may be amplified or minimised by camera technique according to their purposes; and variations in camera height are a ready source of bias in this respect (Mandell and Shaw, 1973; McCain, Chilberg and Wakshlag, 1977).

A lateral variation between full face and profile shots has a coded significance which is less easy to decipher. Intuition may suggest that the direct to camera shot should connote directness of approach and its attendant qualities of authority and reliability; the present data, however, suggest otherwise. If a general rule for such effects is to be argued from this evidence it should perhaps be as follows: that a full face shot suggests less expertise than a profile shot since in popular broadcasting those who address the camera directly are typically the reporters and link men, who transmit the news rather than initiate it. The expert on the other hand is more often seen either in interview or in discussion, and thus in profile. Unless the speaker may be assumed an expert on some other basis—which, as suggested in Experiment 2, the conventional TV reporter is not—the

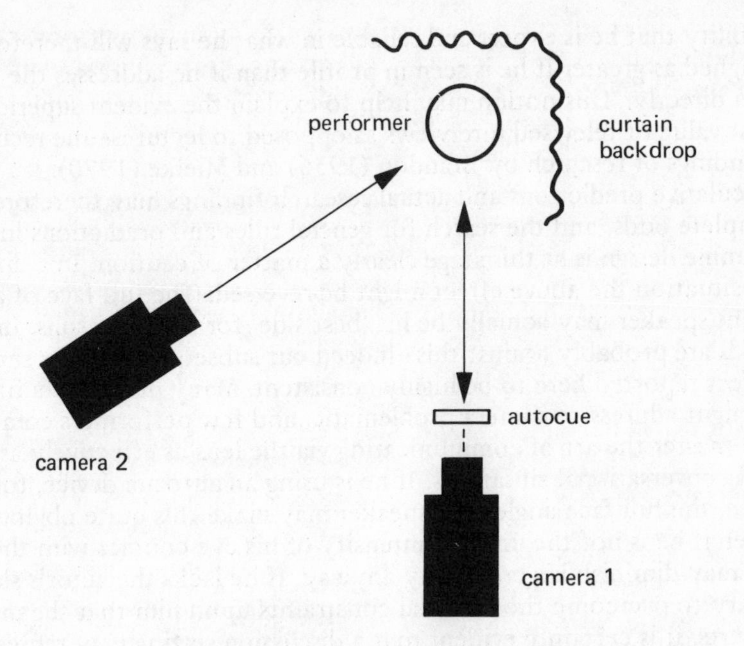

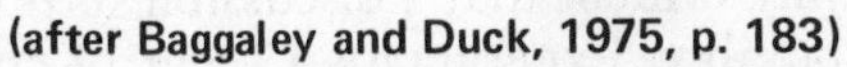

Figure 2.5 Studio floor plan for direct/profile recordings — Experiment 3

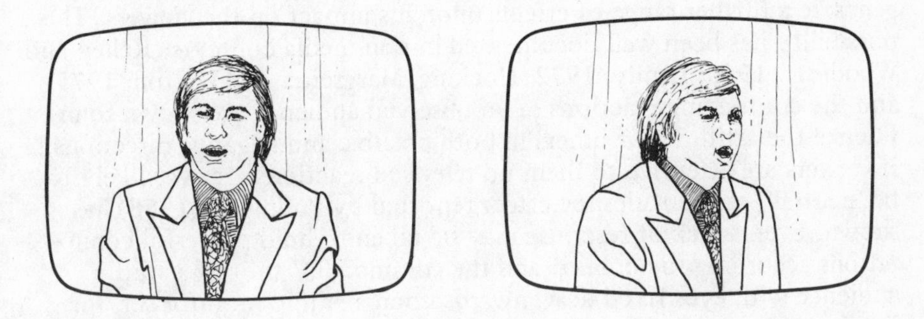

Figure 2.6 Direct address: Camera 1 — Experiment 3

Figure 2.7 Profile address Camera 2 — Experiment 3

probability that he is expert and reliable in what he says will therefore be weighed as greater if he is seen in profile than if he addresses the camera directly. This notion may help to explain the evident superior interest value of televised interviews as opposed to lectures—the recurrent findings of research by Brandon (1956) and Mielke (1970).

Speculative predictions and actual research findings may therefore be at complete odds, and the search for general rules and predictions in TV programme design is at this stage clearly a matter of caution. In a different situation the above effect might be reversed. The full face of a different speaker may actually be his 'best side' for many reasons. In fact the odds are probably against this—indeed our subsequent results show the effect reported here to be highly consistent. Many performers find the straight address to camera problematic, and few performers completely master the art of communicating via the lens as effectively as in normal conversational situations. If he is using an autocue device, for instance, the full face angle on a speaker may make this quite obvious; and even if he is not the unusual intensity of his eye contact with the viewer may diminish his credibility anyway. If he lacks the actor's skill necessary to overcome the artificial constraints upon him that the medium exerts, it is certainly evident that a discussion setting may represent the speaker more favourably. Experiments 13 and 14 consider the influence of camera eye contact on viewers' reactions in more detail.

Experiment 4: Audience shots

In the 'indirect' situation in which the lecturer or presenter is addressing a televised audience, information regarding their reaction to him may generate a further range of effects upon his impact on the viewers. This possibility has been well documented in non-media contexts (Kelley and Woodruff, 1956; Landy, 1972; Hocking, Margreiter and Hylton, 1975) and the contrasting reactions of an observed audience are shown to influence the attitudes of others in both positive and negative directions. If viewers are attentive to them no televised reactions are ever likely to be 'neutral', as the Kuleshov effect reported by Pudovkin (1958) has shown. Even a lack of response may be taken to hold powerful connotations according to context: and the common TV shot of a studio audience with eyes fixed heavenwards upon a studio monitor set, for instance, suggests quite unmistakeably their preference for watching themselves rather than the person before them!

In order to increase its general entertainment value producers commonly add pre-recorded audience applause or laughter to the soundtrack of a presentation similarly; though visual shots of an audience are usually presented for reasons of production variety rather than in the more deliberate attempt to bias the viewers' reaction to a performer more subtly. In order to test whether the visual practice may have unintended influences upon viewers' attitudes, however, the effects of various pre-recorded audience reaction shots in a televised lecture presentation were now investigated.

Procedure

A three and a half minute extract from a televised lecture on welfare economics was selected, during which the lecturer was seen in medium close up at a lectern working from notes, and in near and long shots at the blackboard. The general visual detail was varied using a single camera with zoom lens. The extract was copied onto two videotapes. At the same point in each of the two recordings, shots of an audience were presented, prepared independently with the help of a group of student actors and edited into the lecture to give the illusion that lecturer and audience had been together at the time of recording. No audience sound was presented, and while audience reactions were seen the soundtrack of the lecture continued uninterrupted. The reaction shots were of two types: positive vs. negative (i.e. approving vs. disapproving. Positive reactions were edited into one tape, negative reactions into the other. Each of the two versions was presented to one of two student groups as previously (N = 13, 12) and their reactions to the lecturer were invited on 16 semantic differential scales. Further detail of the experimental design is given by Baggaley and Duck (1975).

Results

T-tests on each scale revealed highly significant, if unsurprising, effects on the lecturer's perceived popularity and interest value ($P < 0.01$ in each), the videotaped condition into which the positive audience reactions had been inserted receiving the more favourable assessments. These results serve to affirm the illusions of continuity and context created by the editing technique. In addition, however, effects on further scales indicate the apparent influence of the editing procedure on reactions to the speaker's message. For in the negative condition she was seen as significantly more CONFUSING ($P < 0.01$), and more SHALLOW and less EXPERT ($P < 0.05$ in each) than in the positive condition; and it should here be re-emphasised that in both conditions the lecturer's performance was held constant, only the reaction shots being varied.

Discussion

The edited insertion of varying audience reaction shots affected the impact of a lecturer's message as well as more personal qualities of popularity and interest value.

The powerful persuasiveness of this technique as a propagandist device is evident: the insertion of favourable audience reactions may substantially increase the motivational value of a recorded lecture even though the reactions are simulated by actors and inserted by a film or videotape editor. Although in an educational production the deliberate insertion of negative reactions would naturally be unacceptable on ethical grounds, the inclusion of favourable reaction shots in order to improve a lecturer's perceived expertise, profundity and clarity as above would hardly be less ethical than any of the other techniques which producers employ to increase the effectiveness of their presentation. Editing, dubbing, keying and superimposition techniques all similarly aim at illusion; and a pre-recorded selection of standard audience reaction shots could prove a valuable addition to the educational producer's stockpile!

In view of these findings, producers should certainly guard against the random use of audience shots selected solely in order to vary the content and pace of presentation. The possible negative influences on the viewers' reaction to a speaker may outweigh any benefits to be derived.

Experiment 5: Interviewer shots

Much illusion in the visual media is created by one or another form of editing procedure as in Experiment 4. However, when the need for editing arises for reasons beyond the producer's control, either due to faulty camera work or limitations in working conditions, then the edited effects may be imperfect. In location filming, shots are typically recorded out of sequence for convenience and subsequently edited together. In situations where only a single camera is used the illusion of a multi-camera situation may be created—even within the presentation of, for example, a continuous interview—by:

1 Training the camera on the interviewee throughout the recording session.

2 Re-recording the interviewer's questions and facial reactions ('cutaway' shots) when the interview is over.

3 Editing the sound and vision together in a manner that maintains the continuity as best possible.

To the viewer unaware of this technique the illusion of continuity may be effective enough, though the occasional inappropriate nod or smirk on the intervewer's part may cause certain bewilderment; and, as Experiment 4 has suggested, information offered by visual reaction shots of other types may have quite substantial effects generally. Accordingly, an experiment was conducted to determine whether viewers' reactions may be influenced by the edited 'cutaway' technique conventional in TV interview situations.

Procedure

A three and a half minute extract from a televised interview was selected. The interviewee, a poet, was seen answering questions about his work from a colleague familiar with it and able to maintain a continuous and easy flow of impromptu conversation. The two sat in opposing armchairs, with a camera permanently trained on each one. While the interviewee was seen in a variety of close and long shots, shots of the interviewer were fixed at medium close up from head to waist level. As in the previous experiments two versions of the presentation were prepared. The first was straight and unedited. By an editing process, however, all shots of the interviewer were replaced in the second version by material recorded after the actual interview was over, in which he repeated his questions and gave reactions to imaginary replies. Differing from the first version solely in the presentation of the interviewer, the second was thus effectively prepared according to the technique described above in which a single camera is used and yet the impression of several manufactured.

Viewers' impressions of (a) the interviewer and (b) the interviewee—two separate groups in each case—were obtained on 16 semantic differential scales. The subject numbers per group were (a) 13 and 12 and (b) 10 and 10.

Results

The style of analysis for (a) was as previously. In the edited version of the recording the interviewer was seen as significantly more TENSE ($P < 0.01$), not surprising in view of the instructions given him to pose his questions and to simulate reactions *in vacuo*. He also appeared more SINCERE ($P < 0.05$), an effect which derives from his notably successful effort to perform the role realistically. More interesting in the present context is the finding that, in the unedited version, the interviewer was seen as significantly less STRAIGHTFORWARD (i.e. more CONFUSING: $P < 0.05$) than in the edited version, once again an apparent effect on the impact of his message.

When effects on viewers' reactions to the interviewee were analysed in (b), significant differences between the edited and unedited versions were observed on a surprisingly large number of scales. Whilst there were no differences between the viewers' perceptions of his nervousness or sincerity in the two tapes, the interviewee was seen in the edited version as significantly more PROFOUND ($P < 0.01$), more STRAIGHTFORWARD and EXPERT ($P < 0.05$ in each)—all effects pertaining to his professional worth as a teacher. Other significant effects showed him to come across in the edited version as more PLEASANT, HUMANE and FAIR ($P < 0.01$ in each), more RELIABLE and more STRONG ($P < 0.05$ in each). Without exception, his perceived qualities were enhanced by the editing procedure rather than diminished; and all of these effects were, it will be remembered, caused by the edited variation of the interviewer's shots, while the interviewee's coverage in the two videotaped conditions was identical.

Discussion

(a) The re-recording and edited insertion of the interviewer's role in a discussion increased his perceived tension and sincerity as well as his apparent intelligibility.

Clearly the effects on audience responses of such a manipulation may be as varied as the cutaway shots selected; and further investigations of viewers' reactions to the relationship between two discussants, and to each of them given varying contextual shots of the other, might yield interesting differences of a more subtle nature than observed here. (We merely aim to show that such effects occur.) However, certain inferences regarding the effective use of this particular editing technique in educational and news contexts are nonetheless possible.

Primarily, the need for care in preserving continuity in the use of the edited cutaway technique is indicated. For the interviewer the main problem in re-recording his role is to preserve a natural manner even in the absence of normal feedback from the interviewee. In the present experiment every effort was made to avoid the customary pitfalls of the technique and to create the edited illusion of continuity effectively; though still significant effects were generated of an unpredicted type.

None of the negative effects that might arise from hasty editing were observed on this occasion; and the increase in the interviewer's apparent tension in the edited version is the only effect which on its own might be considered adverse. Yet we may consider it fortunate that this same degree of relative tension was accompanied in this condition by a significant increase in his perceived sincerity. Evidently tension and relaxedness are each ambiguous qualities—see Experiment 17 and Chapter 5—and it may be concluded that, if carefully used, the cutaway technique which affected their perception has a general potential for actually enhancing a presentation's impact, additional to its possible detrimental effects.

The unforeseen role of the technique in heightening the interviewer's apparent intelligibility (as measured from STRAIGHTFORWARD to CONFUSING)indicates several positive production guidelines. One may question, for example, the value of using two cameras and the necessary ancillary apparatus in prerecorded situations where the use of a single camera and simple editing techniques can be shown to produce a more powerful educational format. In view of the programmed learning tenet that educational material should be organised in a logically developing sequence of recognisable steps (cf. Kay, Dodd and Sime, 1968), the straight transmission of an unscripted interview or discussion situation may not necessarily be an effective educational strategy at all. If the interviewee alone is seen the effect is naturally likely to be monotonous. If two cameras are used, however, in the effort to cover both participants, the director cannot always predict the moment at which to transfer from one to the other; and when he is only fractionally late in switching to the participant now speaking the auditory and visual presentations may conflict (see Experiment 2). Even momentary confusions of this sort can be argued to weaken the logical structure of the production; and when more than two cameras are used the problems multiply. By the careful use of one camera and an edited cutaway technique in such situations, producers may add structure to an initially unstructured situation, interviewers may rephrase their questions more succinctly, and dangers of confusion in the system can be reduced. The subtle potential of this technique is further indicated in the second section of the experiment.

(b) The re-recording and edited insertion of the interviewer's role in a discussion substantially enhanced the general impact of the interviewee.

Two possible explanations for this effect may be offered. On the one hand it may stem directly from the manipulation of the interviewer shots: that the two tapes differed appreciably in respect of the latter's performance is evident from section (a) of the experiment. This interpretation is consistent with the results of Experiment 4, and recalls an incident reported to the original authors by a professional TV interviewer following his filmed election conversation with a political candidate. Though the interviewer had taken care to appear totally impartial during filming, the candidate's agent regarded certain interviewer cutaways later inserted as detracting from the interviewee's credibility: and he prevented the film from being screened. The present findings suggest that this intuition may have been sound; for we see that whatever the participants' intentions during filming, both of an interview and of subsequent cutaway material, the edited manipulation of interviewer shots alone may affect viewers' perceptions of both participants to a surprising degree.

A second explanation of these findings lies in a possibility also indicated in the discussion of Experiment 5a: that the editing procedure itself may have succeeded in clarifying the interview's structure, rendering the boundaries between question-answer-question more emphatic than in the unedited version. An improved interview structure may enhance viewing attitudes to interviewer and interviewee alike; and the concerted results of Experiments 5a and 5b thus testify to the vital role of structuring and sequencing in the impact of communications generally as indicated by, for example, Schegloff (1968). In any conversational situation a meaningful structure is imposed as much by the participants' non-verbal behaviour as by their actual verbal exchanges; and each participant in a dialogue is deeply sensitive to this behaviour's underlying significance (Argyle, 1975). In a media situation—where sequence and structure are further modulated by the TV director and his production team—it is only natural that such effects as here reported should arise for reasons of presentation technique as well as performance. Moreover, as these two factors undoubtedly interact, it is possible that both explanations for the present findings are appropriate to an extent, as compatible rationales for different aspects of the findings.

Experiment 6: Combined variables (students)

The results so far have generated a number of suggestions for the producer of, for instance, educational TV and the consistency of visual background effects in particular has been indicated. However, for the practitioner's benefit the next priority is to test the reliability of image effects in a variety of TV situations based as closely upon conventional practice as possible. The fact that the previous experiments have each studied the impact of a specific TV presentation variable in isolation limits their general applicability. In normal production contexts several variables may be manipulated at once, and the effects of individual variables may consequently be modified. We will now attempt to replicate certain of the above results, therefore, in contexts in which variables are simultaneously combined.

Each variable, in addition, needs to be tested in conjunction with different performers and subject matter. When the text of a presentation is overtly persuasive or propagandist, for instance, presentation effects may actually be diminished; for, as McGuire (1972) has shown, reactions to a communication are less likely to be influenced when an individual is aware that a persuasive attempt is being made upon him. The image variables studied previously, therefore, should be checked in the context of this type of subject matter in contrast to the straightforward and factual styles of text we have used hitherto. As a further constraint, variables should be tested over extended periods. The materials used in our earlier experiments have lasted between one and four minutes; yet Argyle and McHenry (1971) have indicated that after five minutes the effects of surface cues to credibility in others fade and are replaced by more substantial bases for judgement. It is possible that visual cues to credibility in the TV situation are similarly shortlived.

A further experiment was therefore conducted, differing from those reported in our earlier series in these respects:

1 Four of the earlier variables likely to be encountered simultaneously in normal production contexts were examined in a more complex interactive manner via a multivariate design.
2 The TV format, performer and subject matter were all different from those studied previously.
3 The text used was an overt attempt at persuasion and was thus calculated to raise viewers' resistance to the techniques used in conveying it.
4 The material lasted over five minutes and substantial presentation effects were not, therefore, predicted.

The experiment—described in preliminary form by Baggaley and Duck (1976)—thus constitutes a stringent test of the previous effects' robustness (see also Baggaley and Duck, 1979).

Procedure

TV material in the style of an appeal was prepared on behalf of a fictitious charitable organisation, the British Vagrancy Trust (BVT). Recorded using actors in Liverpool University's TV studio and on location (e.g. vagrants living rough; vagrants receiving food at a soup kitchen; contextual shots of Liverpool dockland), the material was assemble edited into six videotape versions. In the studio two basic recordings were made, using two cameras placed at different angles to the speaker; the latter enacted the BVT's 'director', using an autocue. One camera recorded him directly facing it ('Direct tape'); the other framed him in profile as though he were addressing an unseen audience or interviewer ('Profile tape'). The two recordings were simultaneous as in Experiment 3, ensuring that all performance variables were held constant (see Figures 2.5 to 2.7). Each was designed for use (a) in its own right as one of the experimental conditions, and (b) as a basis for two more versions created by the edited insertion of location material at identical points.

One such composite was formed by editing into a copy of the 'Direct' tape several long shots of the speaker—also recorded simultaneously with the basic versions–which revealed a sheet of notes held on his knee (Direct + Notes tape) (Figure 2.8). At the same points in the 'Profile' copy of the tape, shots of a second person nodding agreement and looking pensive were inserted (Figure 2.9) creating a Profile + Reaction tape. A fifth and sixth version of the appeal were created by editing into copies of the previous two tapes a number of filmed location shots (e.g. Figures 2.10 to 2.11), again at the same points in each: Direct + Notes + Location and Profile + Reaction + Location versions of the same basic material were afforded respectively. The insertion of location shots served to provide relevant contextual detail comparable with that used in still form in Experiment 2.

Figure 2.8 'BVT' cutaway shot: 'Direct + Notes' – Experiments 6 to 8

Figure 2.9 As 2.8: Interviewer reaction

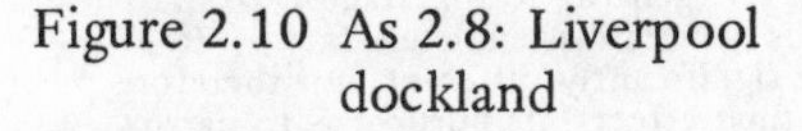

Figure 2.10 As 2.8: Liverpool dockland

Figure 2.11 As 2.8: Vagrant

Further combinations of material would have been theoretically possible (e.g. Direct + Reaction, also Profile + Notes) though these would not have been conditions so likely to arise in production practice, and they are thus of less immediate research interest. As more variable combinations are attempted, moreover, the technological problems of obtaining simultaneous recordings are increased; and the two combinations suggested above would certainly not have been technically possible, given the resources at our disposal, unless material recorded during separate studio performances had been used. In the present design, of course, only one performance is viewed albeit from different viewpoints. The general rationale for the six versions used is indicated via the columns of Table 2.1; the two conditions in each represent a particular level of visual detail, varying from the meagre level (A and B) to an enriched level (E and F). The two rows of the table contain reportive (Direct) and discussive (Profile) formats respectively.

Table 2.1

'British Vagrancy Trust' design:
Experiments 6 to 8

Format	*Levels of contextual detail*		
	(1)	(2)	(3)
	Unrelieved single shot	Level (1) plus detail re performer	Level (2) plus detail re message
Reportive (direct address)	A	C (Cutaways showing notes)	E (Illustrative location shots)
Discussive (profile address)	B	D (Cutaways showing interviewer reactions)	F (Illustrative location shots as in C)

(after Baggaley and Duck, 1976, p.100)

An important feature of the presentation was a plea for money to aid in the Trust's expanding work. It should be stressed that the soundtracks of all six versions of the appeal were identical: verbal content and the performer's speed of delivery, inflection and manner were thus unaffected by the editing manipulations, the latter being in visual production style alone. The text and visual summary of the appeal in its various versions is given in Appendix 2, obtainable from the author; all versions lasted precisely five minutes and sixteen seconds.

Sixty subjects took part in the study, all undergraduate students at Lancaster University. They were divided into six groups each containing equal numbers of each sex. None had seen the actor(s) on the videotape previously. Each subject was shown only one version of the appeal, as in the previous experiments, and was then asked to rate its presenter by means of 16 semantic differential scales.

Results

The analysis was lengthier than in the previous experiments in view of the multivariate design. Its first stage was a series of separate two way analyses of variance on the scores of each attitude scale. This analysis indicates whether any two or more of the six sets of scores are significantly different; it also separates the variances arising between the rows of Table 2.1 (general 'format' factor) from that arising in the comparison of its columns (general 'visual detail' factor). It does not identify which pair of means are significantly different, nor therefore the visual cues necessarily responsible for such effects; its purpose is to narrow the field to particular attitude scales for further investigation.

Viewers' attitudes to the six videotape conditions were significantly differentiated on four of the scales, as shown in Table 2.2: while perceived relaxedness, strength and straightforwardness are each related to the 'detail' factor, perceived expertise is related to 'format' and to an interaction between the factors. Subsequent t-test comparisons of individual videotape conditions on these scales (Table 2.3) revealed the particular presentation variables having differential effects. When the speaker's 'notes' detail was visible he was rated significantly more STRONG (A vs. C comparison: $P < 0.05$) than when they were not. When location detail was added he was rated significantly more STRAIGHTFORWARD

(C vs. E and D vs. F comparisons: $P < 0.05$ in each). No significant attitude differences were attributed to the basic camera angle variable (A vs. B) nor to the presence or absence of interviewer reactions (B vs. D): we must therefore assume the above 'expertise' and 'relaxedness' effects to arise from an unspecified interaction of cues. On all four scales, however, the speaker was rated most favourably in condition F (Profile + Reaction + Location shots), and significantly so at the 5 per cent level or higher in comparison with condition B (Profile alone). On all but the 'strength' scale, F was rated significantly more favourably than C (Direct + Notes), the object of the least favourable ratings on these scales: the three CF differences were each significant at the 1 per cent level. C was also regarded as more CONFUSING than E (Direct + Notes + Location) at the 5 per cent level.

Discussion

Added location detail and evidence of a speaker's use of notes affected his straightforwardness and strength respectively, as perceived by students.

The previous evidence that the addition of general visual detail enhances students' attitudes to TV material is thus supported. The direction of its effects has remained constant despite changes in programme context, performer and subject matter. The 'use of notes' visual cue is shown to have an effect upon subjects' attitudes in the conditions in which it was manipulated independently. The rating of the speaker with notes as more STRONG than when seen without them, however, is contrary not only to the general low rating of condition C on other scales but also to the direction of the previous 'notes/no notes' comparison in Experiment 1. The present result is possibly due to an unforeseen and therefore uncontrolled visual cue arising from the long shot technique used on this occasion (see Figure 2.8). It relates to certain behavioural effects reported under Experiment 18.

In relation to the effects of general visual detail at least, the prediction stemming from McGuire's work (1972)—that subjects may gain resistance to presentation effects when viewing TV material obviously intended to be persuasive—is not generally upheld. Certain effects noted previously (i.e. of camera angle and interviewer reaction) are not evident on this occasion, and it is not within the present experiment's scope to explain this. Not having been found to exert a differential effect in the conditions in which they alone were manipulated (A vs. B, and B vs. D respectively) they cannot be assumed to have any impact in other conditions. It is nonetheless clear that, despite the overt appeal format of the TV material, variations of visual detail continue to exert a marked effect on viewers' responses as in the previous experiments.

The hypothesis derived from work by Argyle and McHenry (1971)—that during an extended (five minute) period visual presentation effects will deteriorate—is no more successful than that à la McGuire above. Variations in general visual detail at least have exerted a notable influence over the longer period despite the methodological constraints imposed upon them in the experiment generally. Of course, certain variables have

Table 2.2

ANOVAs/2: 'BVT' conditions (students) – Experiment 6

Scale	*Mean scores*						*F scores*		
	A	B	C	D	E	F	Format	Detail	Interaction
RELAXED	5.3	6.2	6.5	5.3	5.7	4.3	2.21	3.27*	3.72*
STRONG	5.1	4.8	3.8	4.1	4.0	3.4	0.56	6.20†	0.56
S'FORWARD	4.6	5.0	6.0	5.3	4.4	3.4	1.91	5.07†	0.26
EXPERT	5.4	5.7	6.8	4.7	6.4	4.0	10.83†	0.57	4.03*
df							1,54	2,54	2,54

(NB Lower means denote more positive ratings; significance levels: *5%, †1%)

Table 2.3

Significance levels of between condition differences – Experiment 6

Scale	(notes) AC %	AD %	AE %	AF %	BF %	CF %	(location) DF %	EF %	CD %	(location) CE %	DE %
RELAXED					1	1					
STRONG	5	5	5	1	5						
S'FORWARD				5	5	1	5			5	
EXPERT					5	1		1	1		5

(NB Only those conditions significantly differentiated on at least one of the four significant attitude scales are tabulated; *after Baggaley and Duck, 1979, p.9*) been diminished in the present context as we have noted: these may well have been reduced by the viewers' greater need for more general visual variation in this situation—a possibility that is consistent with Coldevin's observation (1978a) that the impact of a single visual background image upon interest arousal may diminish over longer periods unless sustained by further variation. But whether a presentation is relatively long or short the impact of general visual technique is clearly maintained in a variety of natural formats; and it is likely that a longer presentation period merely causes viewers' visual responsiveness to shift between variables rather than to be reduced generally.

Further investigations may identify more specific relationships between presentation effects and the types and duration of TV material from which they arise. The present experiment suffices to re-emphasise the need for TV producers of educational material particularly to aim for a varied style of presentation over extended periods, and yet to attempt to anticipate the subtle, combined effects of presentation that may occur in the process. It is a dilemma that researchers may help to resolve by continued experimentation into the sources of image effects and the possible rules underlying them.

Experiment 7: Combined variables (children)

A factor as yet not directly acknowledged in these experiments is the viewer himself. Our subjects so far have all been undergraduate students, who may feasibly bring a wealth of intelligence and sophistication to the perception of TV imagery that other subjects do not possess. The rules underlying image effects upon audience attitudes may clearly vary greatly from one viewer to the next, as studies using the Kellian repertory grid technique by Baggaley and Duck (1976) have indicated. Are presentation effects due to a high degree of visual intelligence, or conversely are they increased at lesser levels of sophistication? In order to find out, Experiment 6 was now replicated using schoolchildren.

Procedure

The material previously prepared for Experiment 6 was presented to a total of 74 schoolchildren, members of the Fifth and lower Sixth forms at Merseyside secondary and comprehensive schools: their ages were from fourteen to seventeen years. As before, subjects were tested in small groups and each saw one only of the six videotape versions. Testing continued until each version had been viewed by at least ten subjects. Following the TV presentation the subjects recorded their assessments of the performer on the same 16 rating scales as used in the previous experiment, though in reversed order. In all other respects the procedure and analysis of the two experiments were identical. (NB: Reversal of scales in semantic differential testing does not, according to Warr and Knapper (1968, p.86), have any significant effect on results; in this context, therefore, it was cautious though not apparently strictly necessary.)

Table 2.4

ANOVAs/2: 'BVT' conditions (children) – Experiment 7

Scale	*Mean scores*						*F scores*		
	A	B	C	D	E	F	Format	Detail	Interaction
RELAXED	4.1	4.2	5.9	4.5	3.6	3.3	1.69	5.77†	1.17
STRONG	3.4	4.1	4.9	4.0	3.5	2.7	1.01	4.78*	2.14
PERSUASIVE	2.9	5.1	5.1	3.3	3.3	3.3	0.09	1.51	6.80†
HUMANE	3.4	2.0	3.4	2.4	1.8	1.5	7.73†	5.26†	0.96
PLEASANT	3.6	2.1	3.9	3.9	3.3	2.3	5.89*	3.98*	1.46
S'FORWARD	3.1	1.9	4.5	3.0	1.3	1.7	4.33*	12.54†	2.66
FRIENDLY	2.3	2.4	3.6	3.7	3.4	2.2	0.74	4.49*	1.36
PROFOUND	3.4	5.1	4.6	3.9	3.1	3.7	2.32	2.60	3.54*
INTERESTING	4.1	3.5	6.1	4.5	3.3	2.4	6.39*	12.53†	0.50
POPULAR	4.7	4.0	4.6	3.8	3.7	2.8	7.44†	5.22†	0.05
EXPERT	4.0	4.3	6.1	2.9	3.0	2.8	7.85†	6.49†	8.53†
FAIR	4.2	2.2	3.0	3.0	1.6	1.6	4.34*	9.03†	4.12*
SINCERE	2.4	2.5	3.4	2.3	2.0	1.5	3.02	4.87*	1.43
df							1,68	2,68	2,68

(NB Lower means denote more positive ratings; significance levels: *5%, †1%)

Table 2.5

Significance levels of between condition differences – Experiment 7

	(angle)	(notes)					(reaction)				(location)			(location)
Scale	AB %	AC %	AD %	AE %	AF %	BC %	BD %	BE %	BF %	CD %	CE %	CF %	DE %	DF %
RELAXED		5				5					1	1		
STRONG		5							5		5	1		5
PERSUASIVE	1	1					5	5	5	5	5	5		
HUMANE	5			1	1	5					1	1		
PLEASANT	5				5	1	1	5				1		1
S'FORWARD		5		1	5	1				5	1	1	1	5
FRIENDLY		5	5				5					5		5
PROFOUND	1	5					5	1	5		5			
INTERESTING		1			5	1				5	1	1		1
POPULAR				5	1				5			1		5
EXPERT		1			5	1	5	5	5	1	1	1		
FAIR	1	5	5	1	1						5	5	5	5
SINCERE		5							5	5	1	1		

(NB Significant differences thus occur between each pair of conditions except EF)

Results

Two way analyses of variance on each of the 16 attitude scales again revealed significant differences between audience ratings of the performer in the different conditions—but on a surprising 13/16 scales as opposed to the mere 4/16 thus affected in Experiment 6. The latter scales (relating to perceived strength, straightforwardness, expertise and relaxedness) were significantly differentiated once again; and on all of the 13 scales except for those measuring persuasiveness and profundity the effect was significantly related to the between columns comparison (Table 2.1), and thus to the general 'detail' factor. The more detail and variety of shots in the presentation, the more favourable the children's reactions to the actual performer. Table 2.4 presents the means and ANOVA/2 results on the 13 significant scales.

The analysis also suggests that differences between the direct and profile formats, arising from the ACE vs. BDF comparison of rows in Table 2.1, were influential too. Seven of the 13 significant scales yielded effects related to this factor, including that observed on the expertise scale as in Experiments 3 and 6. On five of the scales this result is confirmed by straight t-test comparisons of the scores relating to conditions A and B—those conditions in which the camera angle variable alone was manipulated. In the Profile condition (B) the speaker was regarded as significantly more PLEASANT and HUMANE ($P < 0.05$ in each) and FAIR ($P < 0.01$) than in the Direct condition. But the angle cue also evidently proved ambiguous to these subjects, for the Profile condition was simultaneously regarded as less PERSUASIVE and PROFOUND ($P < 0.01$ in each) than the Direct. A virtually identical mixed response was found to the 'interviewer reaction' cue, as measured by the comparison of B vs. D ratings: when the reactions were shown, the speaker was regarded as significantly more PERSUASIVE, PROFOUND and EXPERT ($P < 0.05$ in each) than when they were not, but also less PLEASANT ($P < 0.01$) and FRIENDLY ($P < 0.05$). A strong response to the 'use of notes' variable (A vs. C) quite unequivocally favoured the 'no notes' condition: on nine of the scales the speaker was regarded more favourably without notes than with them. The favourable influence of added location detail upon attitudes to the speaker was confirmed in the comparison of conditions C and E (on ten scales) and conditions D and F (on seven scales). The total and level of all significant differences between conditions are indicated in Table 2.5. The only scales on which ratings were not significantly differentiated by presentation were those concerning perceived spontaneity, honesty and reliability.

Discussion

Added location detail, evidence of a speaker's use of notes, interviewer reactions and variable camera angle had often ambiguous effects on his qualities as perceived by schoolchildren.

The replication of location detail and 'notes' effects in this experiment adds weight to the results reported earlier in the chapter. In comparison with the student subjects the children certainly exhibit a dramatically greater responsiveness to visual cues: significant effects are attributed to each one of the image variables manipulated.

The impact of added visual detail particularly is confirmed by the subjects' relative preference for the conditions containing location shots. The 'use of notes' cue, largely overwhelmed when examined in conjunction with other variables in Experiment 6, is influential on a total of nine scales and in the direction confirming the original results of Experiment 1. Responses to the camera angle and 'interviewer reaction' variables suggest two notions that we shall pursue in later chapters. The mixed positive and negative effects of these cues indicate firstly that an audience may be ambivalent in its responses to individual cues, possibly in view of the independent effects of separate aspects of the cue upon it (see Experiment 14 and Chapter 4). A single visual image will thus contain more than one 'seme' or unit of perceived meaning (Fiske and Hartley, 1978). The second possibility, investigated in Chapter 5, is that individual attitude scales may represent different clusters of meaning, independent dimensions of attitude underlying subjects' responses to TV imagery in general.

The relatively heightened responsiveness of the children to visual detail does not, however, necessarily imply that they are more sophisticated in their perceptual abilities than the students tested previously. Both types of subject may be equally proficient perceptually, the difference between them stemming from the children's lesser ability to criticise and possibly to dismiss the impact of presentation variables upon them. This conclusion may thus be viewed in terms of the concept of 'visual literacy' (Arnheim, 1969). The idea that uses of visual imagery by the film and TV media are based on a system of grammatical rules which viewers learn to decode has been discussed in detail by structural and semiotic analysts (cf. Hood, 1975; Fiske and Hartley, 1978). The need to instruct viewers in the extent to which presentation may influence their judgements—thus in the 'decoding' process specifically—was emphasised by Baggaley and Duck (1976); only thus, it was felt, may they learn to resist the possible influence of propagandist visual stylistics, and also to make better use of visual information in news and educational contexts. The present results suggest that such instruction is particularly, though not exclusively, desirable for children. The extent of their responsiveness and their tolerance of visual ambiguity collectively suggests a susceptibility to very subtle visual persuasion easily capable of exploitation.

Experiment 8: Combined variables (clerical staff)

Though suggestive, the difference between the results of Experiment 6 and 7 does not therefore imply that children are more susceptible to presentation effects than adults in general. Within the adult population undergraduate students—as used in Experiment 6—are still a special case; and in order to check whether their responses to visual presentation may be taken to predict those of other adult groups, a final replication of the multivariate experiment was now conducted.

Procedure

The subjects on this occasion were 66 clerical workers, members of the administrative and secretarial staff of Liverpool University answering a request for experimental subjects posted on departmental noticeboards. In all other respects the procedure was identical to that of Experiment 6.

Table 2.6

ANOVAs/2: 'BVT' conditions (clerical staff) – Experiment 8

Scale	*Mean scores*						*F scores*		
	A	B	C	D	E	F	Format	Detail	Interaction
RELAXED	5.5	5.9	5.2	3.3	5.5	3.5	9.98†	5.58†	4.58*
POPULAR	4.0	2.7	4.3	3.1	3.4	3.4	4.57*	0.33	1.15
EXPERT	5.3	3.6	4.2	3.2	4.2	4.0	4.47*	0.88	0.88
df							1,60	2,60	2,60

(NB Lower means denote more positive ratings; significance levels: *5%, †1%)

Table 2.7

Significance levels of between condition differences – Experiment 8

Scale	(angle)	(reaction)				
	AB %	AD %	BD %	CD %	DE %	BC %
RELAXED		1	1	1	1	
POPULAR						5
EXPERT	5	1				

(NB Only those conditions significantly differentiated on at least one of the three significant attitude scales are tabulated)

Results

The extent of presentation effects upon these subjects was of the same order as that observed in Experiment 6. Two way analyses of variance upon each of the attitude scales showed only 3/16 of them to be significantly differentiated by presentation technique. Two of the significant effects, however, were observed on the two scales (perceived relaxedness and expertise) significantly affected by presentation in both of the immediately preceding experiments; the third scale affected was perceived popularity. All three of these effects were significantly related to the 'format' factor (Table 2.6). Thus the performer was rated more RELAXED, EXPERT and POPULAR in the Profile conditions overall than in the Direct conditions. Apparent relaxedness was significantly related to both format and visual detail factors, and to an interaction between them.

The impact of the camera angle variable upon perceived expertise was confirmed by the t-test comparison of conditions A and B ($P < 0.05$). No significant differences between ratings were observed in any of the AC, CE or DF comparisons, indicating that the notes and location shot variables were not on this occasion influential. When the interviewer's reactions were presented, however, the speaker was regarded as significantly more RELAXED than when they were not (BD comparison: $P < 0.01$). As Table 2.6 indicates, condition D was the generally preferred condition. It yielded ratings significantly more favourable than all other conditions except F on the RELAXED scale ($P < 0.01$ in each), as is shown together with the other between condition comparisons in Table 2.7.

Discussion

Added interviewer reactions and variable camera angle affected a speaker's relaxedness and expertise respectively, as perceived by clerical staff.

The significant impact of the format factor upon perceived expertise in this experiment confirms the effect now observed on three previous occasions (Experiments 3, 6 and 7). In each, higher expertise ratings have been associated with 'profile' presentations, either owing to the camera angle variation itself or to its interaction with other variables. A further effect consistently observed in the three multivariate experiments involves the association of relaxedness ratings and the amount of visual detail: in the conditions containing the most detail (particularly when featuring the profile presentation also) perceived relaxedness tends to be significantly greater.

From the multivariate experiments in general it is clear that visual presentation variables exert a significant impact on the attitudes of contrasting types of viewer, and also when varied simultaneously as well as individually as in the more artificial experimental situations examined earlier. Yet subjects may respond to different visual cues in different situations; and different types of subject demonstrate different levels of awareness in responding to the same material. In this experiment it has been the more subtle visual cues of angle and interviewer reaction that have exerted most of the effects observed, rather than the more obvious visual detail cues heeded in particular measure by the children—Experiment 7. Having concluded in that context that the development of 'visual literacy' may be a two stage process—comprising (a) the capacity to draw inferences from non-verbal images, and (b) the critical skill to evaluate them—we may now suggest that the older subjects have probably succeeded in dismissing the more obvious inferences as irrelevant to their task, while failing to do so with certain of the subtle ones.

The evidence that individual image variations may yield conflicting effects within an audience's attitudes suggests that the range of meaningful visual cues simultaneously available for an interactive effect may be more extensive than we have hitherto acknowledged and controlled. We examine this possibility in the next chapter. It has also become apparent that for a fuller understanding of image effects we ought to examine the particular dimensions underlying attitudes to TV material. Evidence for systematic psychological effects linking Experiments 6 to 8 is presented in Chapter 4, and attitude factors common to the research as a whole are indicated in Chapter 5.

3 People, words and music

Of the presentation variables reported in Chapter 2 some are clearly more influential upon viewing response than others. Moreover, certain of the attitude scales presented to our viewing subjects appear more consistently susceptible than others to presentation effects in different conditions. Even the adult groups least affected by visual variation (see Experiment 8) seem unable to resist the suggestive influences of TV presentation on perceptions of a performer's relaxedness and expertise. In the absence of more substantial evidence regarding all important qualities in the TV performer such as these, it has been evident so far that viewers do not necessarily suspend all judgement: they are capable of basing imaginative, although possibly ill advised, speculations about a performer on visual cues that would normally be considered quite inadequate for the purpose. The cues themselves stem not only from particular disturbances of behaviour endemic to the TV situation as in Experiment 5, but also from options open to the production team in representing the situation in image form.

In the present chapter we examine the conditions in which image effects operate in more detail. At no stage have we assumed the types of variable examined hitherto to be the dominant sources of audience impact; the object of the next group of experiments is to demonstrate that, ever present, these are capable of subtle interactions with the other content variables they serve to mediate. Thus the types of people seen on TV, their broad styles of performance, the words they emit, and so on, each quite naturally exert rival claims upon the viewers' attention. We investigate textual and performance variables, together with other production factors, henceforward. In our considerations of each we have borne the possibility of their multiple interaction with others carefully in mind, since only in this way may we begin to predict the order of importance that factors assume in a given viewing situation.

Having established that susceptibility to visual effects is not exclusive to university undergraduates–indeed that the latter seem less vulnerable in this respect than adolescents–we have continued to examine student reactions to TV material henceforward. The decision is partly for reasons of natural convenience in a university research environment, and partly to avoid suggestions that we have selected for our experiments only those subjects most likely to yield the most dramatic results. A deliberate 'devil's advocate' position in relation to visual impact has in effect been adopted since Experiment 6 in which we imposed constraints calculated substantially to reduce it. In the next experiment we continue this practice by investigating an hypothesis assumed in the earlier work: that presentation effects will cease to be decisive in conditions in which the viewers have prior knowledge of the person they are judging.

Experiment 9: Prior knowledge of the performer

The experiment tests the 'maximum ambiguity' hypothesis derived from the work of Scheff (1973)—see Chapter 1. Most production variables of the type discussed earlier have been assumed unlikely to exert more than a marginal effect on audience judgements once the latter can be formed on more substantial bases. One of the most powerful variables to emerge in Chapter 2 was that of televised audience reaction. Viewers' judgements of an economics lecturer were influenced to a highly significant degree by the simulated reactions of a student audience to her. In order to check the extent to which even a highly influential cue such as this may be undermined, we now replicate the earlier Experiment 4 with one added manipulation.

Procedure

The same three and a half minute videotaped performance—one version with positive reactions inserted, the other with negative—was presented in this experiment following a brief televised introduction concerning the performer's lecturing experience. Two introductions in all were used, recorded in a plain studio setting contrasting with the lecture room context of the main extracts. Their substance was as follows.

1. POSITIVE: 'One of the problems in assessing a lecturer's ability is to take into account the extent of his or her experience. In the next extract we see a lecturer with considerable experience giving a course actually based on her own textbook on the subject of basic economics.'
2. NEGATIVE: 'One of the problems . . . (as above) . . . we see a new and inexperienced lecturer tackling an aspect of basic economics.'

When each introduction was edited onto the beginning of one of the lecturing extracts the impression given was that the whole was abstracted from a programme concerning teaching skills.

Four experimental presentations were thus created and each was presented to ten student subjects. One set of subjects saw the version containing positive classroom reactions with positive introduction stating the lecturer to be highly experienced (++); a second saw the same reactions preceeded by the negative statement (+–); and two further sets saw the negative classroom reactions combined with either the positive or negative introduction similarly (–+ and – –). Subjects' ratings of the lecturer were obtained on 16 semantic differential scales in each condition.

Results

The prediction that prior information about a performer will reduce the impact of otherwise powerful visual presentation variables was, to an extent, supported. Two way analyses of variance on the scores relating to each attitude scale permit us to recognise certain effects due to both 'studio introduction' and 'audience reaction' factors individually. (No interactions between them were noted.) From Table 3.1 it is evident that 6 of the 16 rating scales were affected to a significant degree by the introduction variable. Examination of the mean scores on these scales indicates that the direction of the effects was predictable in all six cases; i.e. subjects rated the lecturer more positively (more STRAIGHTFORWARD, STRONG, HONEST, RELIABLE, EXPERT and less NERVOUS) when she was introduced positively. This trend also occurred in most of the remaining scales even though the differences upon them did not reach statistical significance.

Table 3.1

ANOVAs/2: Speaker introduction x audience reaction – Experiment 9

Scale		*Mean scores*				*F scores*		
	(INTRO)	+	+	–	–			Inter-
	(REACTION)	+	–	+	–	Intro	Reaction	action
S'FORWARD		2.2	2.8	3.6	3.4	5.63*	0.23	0.90
STRONG		3.8	3.4	4.9	4.9	9.85†	0.23	0.23
HONESTY		2.1	2.0	3.8	2.8	5.94*	1.12	0.75
RELIABLE		2.1	2.5	3.8	3.7	12.81†	0.14	0.38
INTERESTING		2.7	4.5	4.0	4.8	2.50	6.60*	0.98
POPULAR		2.9	4.2	3.7	4.1	0.89	5.26*	1.47
Not NERVOUS		4.4	3.1	5.7	4.7	6.44*	4.05	0.07
EXPERT		3.0	3.6	5.4	4.1	9.26†	0.54	3.98
df						1,36	1,36	1,36

(NB Lower means denote more positive ratings; significance levels: *5%, †1%)

Whereas in the earlier Experiment 4 five of the attitude scales used were significantly affected by the televised reactions, only the effects which had been most significant on that occasion (i.e. on perceived interest and popularity) were now replicated. The table shows the lecturer to have been rated as most INTERESTING and POPULAR when positive reaction shots were included. (NB The general question asked by this experiment is thus answered without recourse to multiple between condition comparisons as in the previous multivariate analyses.)

Discussion

The provision of prior information about a lecturer reduced the effects of televised audience reactions on attitudes towards her.

This experiment puts the psychological importance of presentation cues into the maximum ambiguity perspective previously assumed. It indicates that their overall effects are likely to be more pronounced on judgements of persons, issues and events of which nothing is already known or believed. Since much that is presented on educational TV particularly fits this description, such effects are nonetheless likely to be important in that context. The issues and events communicated via TV news are generally novel similarly; and, according to Birt and Jay 1975/6), the manner in which they are actually presented is also typically confusing–a further reason to suppose that presentation effects may flourish in that context.

Birt and Jay, both TV practitioners, refer to a common 'bias against understanding' due to the tendency of TV journalists to focus attention less upon the underlying causes and meanings of news events than on the daily 'hot diet' of their effects. Of course, whenever attention towards the narrative substance of news and opinion is obstructed we may naturally suppose that the impact of superficial image factors will accrue. Production factors may combine with others to unconsciously bias the viewers' understanding in a variety of ways: and, as Chapter 4 onward indicates, such effects are not necessarily restricted to conditions of simple unfamiliarity. For as the persons who present the news become familiar to their audience numerous stereotypic criteria for judging them are likely to develop. Image factors such as we have hitherto studied may cease at that stage to be influential; and, as we shall see from later experiments, personality and performance factors may take their place in the viewers' regard.

It will be noted that the emphases placed in the studio introductions in this experiment upon credibility characteristics did not override the televised audience reaction effects observed in Experiment 4 altogether. Viewers are evidently discriminating in their use of the overall evidence about a performer, and on this occasion evaluated the lecturer's popularity and interest value, and her more general qualities, on separate bases. Indeed no interactive effects between the separate sources of evidence were found. The operation of independent criteria for the assessment of TV performers begins to be glimpsed; and their combined impact upon viewers' attitudes is examined in Chapter 5.

Experiment 10: Non-verbal content vs. verbal

By holding performance and textual variables constant across conditions we have been able to demonstrate that non-verbal production criteria can exert effects upon viewers' attitudes in their own right. We conclude that in ambiguous situations the visual images of TV can affect assessments of, for instance, the general credibility of the verbal information they accompany. Our own data do not indicate further effects of visual presentation on, e.g. the retention and comprehensibility of verbal information; though Findahl and Hoijer (1976, 1977) report general findings to this effect (see Chapter 4, last section). An independent test of the same hypothesis is now conducted.

Procedure

Experimental conditions were designed in which the usual paradigm was inverted and both presentation and performance stylistics were held constant while verbal content was varied. To vary a performer's subject matter without altering any aspect of his delivery is, of course, highly problematic; but the problem was solved thus. A Venezualan actor took part in a videotaped studio interview as the representative of an undisclosed political party. Two supposed translations of his verbal content were prepared in English, one listing the achievements of his own party

Table 3.2

ANOVAs/2: Presentation mode x verbal content

Experiment 10

Scale	*Mean scores*				*F scores*		
(MODE)	Non-TV		TV				Inter-
(VERBAL)	+	−	+	−	Mode	Verbal	action
RELIABLE	5.5	4.6	4.9	3.4	5.59*	9.93†	0.62
HUMANE	3.9	5.7	3.6	3.6	5.53*	3.11	3.11
SHARP	4.4	3.0	2.6	3.8	0.75	0.03	5.05*
POSITIVE	4.5	3.0	2.5	4.4	0.23	0.10	7.46†
SINCERE	4.7	2.6	3.9	3.0	0.16	9.06†	1.45
MODEST	5.1	5.2	4.5	3.8	4.47*	0.40	0.72
PLEASANT	3.7	5.0	4.1	3.2	2.73	0.22	6.74*
df					1,36	1,36	1,36

(NB Lower means denote more positive ratings; significance levels: *5%, †1%)

Table 3.3

Significance levels of between condition differences

Experiment 10

Scale	*Non-TV* + vs − %	*TV* + vs − %
RELIABLE		5
HUMANE	5	
SHARP		
POSITIVE	5	5
SINCERE	1	
MODEST		
PLEASANT	5	

(NB Only those differences relevant to the experimental hypothesis are tabulated) ('positive' version) and the other setting forward the same details point for point as the failures of a rival party ('negative' version). The texts of this, as of other experiments, are available in Appendix 1 obtainable from the author.

Each text was then presented as a paper handout to a separate group of students (N = 10, 10) and the unknown person 'likely to have written it' was rated on 15 attitude scales in the normal manner. It was next added via rolling subtitles to a copy of the one and a half minute TV performance by the actor. The two videotape versions thus created were presented to two new student groups composed as the first two groups for ratings of the speaker on the same scales. None of the subjects understood Spanish and in the TV conditions they were therefore obliged to read the subtitles in order to understand the interview. In each of the four conditions a preliminary test of the subtitles' legibility was conducted, using independent material, and was successful with all subjects to a distance of approximately ten feet. Instructions were given for the subjects to ensure that they followed the subtitles throughout the interview.

Results

Fifteen two way analyses of variance were conducted on the data, one for each scale; the results and the mean scores of each group are summarised in Table 3.2. Significant effects are found due to each of the experimental factors in turn; verbal content (positive vs. negative) and mode of presentation (TV vs. non-TV). On certain scales effects are related to interactions between the factors. In the TV versions as a whole the performer was judged more RELIABLE, HUMANE and MODEST than were the corresponding paper texts. Multiple comparisons of scores on the seven scales significantly affected in general (Table 3.3) indicate that, in the non-TV mode, the 'positive' content was rated significantly more HUMANE and PLEASANT than the 'negative' content ($P < 0.05$ in each) but less SINCERE ($P < 0.01$) and more NEGATIVE ($P < 0.05$)! Significant differences between the two TV conditions were far less in evidence, despite the instruction to subjects to concentrate on the verbal subtitles: in the 'positive TV' condition the performer was rated (as expected) significantly more POSITIVE than in the 'negative TV' condition, but less RELIABLE ($P < 0.05$ in each).

Discussion

Mediation via TV significantly reduced the differential impact of two contrasting verbal communications.

The general comparisons of TV and non-TV judgements indicate an overall preference for the TV conditions regardless of their supposed message; and, in these conditions, the range of perceived contrasts between the two messages *per se* is approximately halved. The effects of verbal content and non-verbal performance characteristics in the TV context have apparently interacted: and it is probable that the non-verbal impact of the presentation diverted attention from the verbal meanings conveyed. Research by Perry and Williams (1978) into the distracting effects of performance style in TV lecture situations supports this possibility. At the very least it may be inferred that when non-verbal and verbal content are judged simultaneously, as in the standard TV situation, cues perceived in either may reinforce one another or conflict—a possibility similar to the independent channel hypothesis indicated in the discussion of Experiment 2. Similar evidence linking lack of visual-verbal agreement in a news item with impaired news recall is reported by Findahl and Hoijer (1976).

For clear evidence regarding types of conflict between independent attitude cues, however, we must apply a different analytic methodology to that used here: we consider this problem in Chapter 5. But conflict and doubt within viewers' reactions to the 'politician' investigated in this experiment are already made only too clear. The association in the non-TV condition of a positive text and more NEGATIVE ratings indicates that a performer's claims of positive achievement are not necessarily accepted as evidence of same, in political propaganda contexts at least. The same conclusion may be drawn from the ratings in the TV conditions of the performer with 'positive' verbal content as being significantly more POSITIVE but less RELIABLE than when presented with the 'negative' content. Conflicts of precisely this type within popular attitudes to political figures have been noted previously by Osgood et al. (1957) and by Blumler and McQuail (1968).

The purpose of experimental work in this field is not merely to demonstrate that such conflicts may occur but also to identify their determinants, so that they may be controlled and hopefully avoided in applied communication contexts. Following Experiment 2, for example, it was suggested that the deliberate diversion, by production strategy, of viewers' attention to a performer's less favourable visual qualities may increase their awareness of his preferable vocal qualities. We now investigate the evidence which led to this conclusion.

Experiment 11: Visual performance vs. vocal

The earlier Experiment 2, comparing the effects on attitudes to a TV newsreader of pictorial and plain background images, yielded two hypotheses:

1 That, as generally supposed by TV producers, an increase in visual detail creates a more enlivened presentation to which the viewers are better disposed overall.

2 That the speaker's vocal qualities were, on this occasion at least, judged more favourably than his visual appearance; and that his perceived credibility rose when the viewers' visual attention to him was distracted.

Experiments 6 and 7 have since supported the first possibility, indicating the responsiveness to visual detail of children in particular. Alternative hypotheses in these matters may, of course, be fully compatible; and the present experiment was designed to test whether the second possibility, regarding differences between the performer's visual and vocal qualities, was to any extent tenable.

Procedure

The same videotaped performance as rated in the original Experiment 2—a seventy second news type item concerning an archeological dig—was reassessed by new viewing groups (students as previously). One group received the sound plus vision presentation in which a plain background was used; the second saw the same presentation minus the sound; and the third heard the sound only (picture contrast on the TV monitor set turned to black). Each of the three conditions was rated on the same 13 adjectival scales as previously used, by ten subjects. If the second (independent channel) hypothesis is at all tenable, audience ratings of the performer's vocal qualities should be more favourable than those of his visual qualities (Prediction 1); and, since the hypothesis suggests that the earlier ratings of the performer in the 'plain' condition were based to a greater extent on judgements of his visual appearance than of his voice, differences between the ratings of vision only and sound plus vision conditions on this occasion should be minimal (Prediction 2).

Results

The first of these predictions was convincingly upheld, even via the stringent two tailed statistical criterion normally reserved for a non-directional hypothesis. (Though strictly entitled to use the more lenient one tailed criterion here, we have retained the two tailed one for uniformity.) Following one way analyses of variance on the 13 scales (Table 3.4) multiple between condition t-tests were conducted on the four scales significantly affected. The comparisons of sound only and vision only ratings revealed a significant preference by subjects for the former condition on each of these scales (Table 3.5); a further five scales showed the same trend albeit at a non-significant level. Between the sound plus vision and vision only conditions fewer differences were noted—on 2/13 scales only. On balance, therefore, the independent channel hypothesis is supported. (Note, however, that the judgements of sound plus vision and sound only conditions were statistically indistinguishable—a result which tends to detract from the hypothesis' overall strength.)

Discussion

A performer's visual and vocal qualities were rated differentially, in directions predicted following Experiment 2.

The results are not decisive though they do tentatively support the notion that audience responsiveness is not likely to be affected by the simple amount of visual detail in a presentation alone, but that 'the insertion of keyed visual detail may serve to distract an audience's attention from the speaker's visual presentation, leaving his auditory channel as the main communicator' (Baggaley and Duck, 1975). Immediate implications are apparent on this basis for the educational TV producer. Just as he may choose to enhance the visual detail in a production in order to reduce the possible impact of a poor visual performance (see Experiment 2) so we may now suggest that he take care to reduce visual detail in the presentation of a performer who is effective visually but poor vocally!

The interaction of separate production and performance factors is thus indicated once again. Clearly the effects of visual and verbal treatments in a TV presentation—whether independently gauged or in combination—are not matters about which one may yet safely generalise or predict. Travers (1964, 1966) and Coldevin (1975a) each report findings quite opposed to those cited above: both researchers found a sound only

Table 3.4

ANOVAs/1: Aural vs. visual presentations

Experiment 11

Scale	*Mean scores* Sound	Vision	(S + V)	*F score*
PLEASANT	2.3	3.5	2.8	4.05*
BELIEVING	2.4	4.0	2.6	5.23*
FAIR	2.7	3.7	2.4	5.09*
SINCERE	2.1	3.6	2.9	3.78*
df				2,27

(NB Lower means denote more positive ratings; *5% significance level)

Table 3.5

Significance levels of between condition differences

Experiment 11

Scale	Sound vs. vision %	Vision vs. (S + V) %
PLEASANT	1	
BELIEVING	1	1
FAIR	5	1
SINCERE	1	

(NB No significant differences occur between sound and (S + V) conditions) teaching presentation less favourable (in retention terms) than either a sound plus vision or vision only presentation. As Coldevin points out, however, guidelines for production practice in educational TV must ultimately depend more upon subtle questions of the way in which different information channels are used than on the availability of the channels *per se.* Clearly, the aural content of the moment interacts—or competes, according to the efficiency of production design—with other image factors. But we may nonetheless anticipate that identification via research of the separate communication channels capable of influencing audience attitudes will ultimately aid producers to use the presentation techniques at their disposal more purposefully, in deliberately steering the audience's attention between channels for particular effects.

There is no reason to doubt the possible impact of still finer levels of image variation than we have studied. Kreimer (1974), for example, has suggested that TV's aural messages alone comprise at least three levels of meaning interacting with each other in ways which may both jeopardise and distort the educational effects intended. Thus verbal and paraverbal factors, e.g. pitch and intonation, may each contribute to the overall aural impact. At the more subtle levels of variation such as these the options for control over unintended effects are naturally reduced; and the most practical course for TV producers to take in this connection may thus be to elaborate the visual properties of a presentation when they sense that its aural qualities may be lacking. We have investigated some of the options conventionally open to TV news producers in this respect in Experiment 12.

In keeping with the 'devil's advocate' policy adopted in the design of these experiments, however, we have also acknowledged a further possibility in Experiment 12: that a TV performance which is proficient in all respects, aural and visual, may of itself reduce the impact of production factors whatever their potential otherwise. If production factors may compete with and ultimately dominate performance factors, so presumably may a highly proficient performance dominate production. In order to create a situation, as in Experiment 9, in which only the most coercive of production effects are indicated, we therefore now compare the differential impact of six conventional TV news formats on judgements of a professional newsreader.

Experiment 12: TV news report formats

For technical as well as artistic reasons, TV news formats conventionally feature a variety of aural as well as visual contrasts which may individually be predicted to have differential effects on audience attitudes. Selected contrasts are built into the conditions compared below.

Procedure

Six versions of a news type item (sixty-five seconds duration) were prepared using simultaneous recordings of a single studio performance. Three immediately adjacent cameras were used, sited approximately 12 feet from the professional BBC presenter, a local radio performer unknown to the subjects in a TV context. The latter read from a desk script, looking up occasionally to continue reading from an autocue transcript projected at the lens of the central camera. At this distance the slight displacement of his gaze from the outer two cameras was not apparent.

While the first camera presented him against a plain studio background, as in Experiments 2 and 11, the images from the second and third cameras were each fused via an electronic 'keying' procedure with relevant and irrelevant picture background images respectively. Three separate visual conditions were thus created. The vocal soundtrack was then used separately in the creation of a further three formats as used in standard TV situations when externally produced material is transmitted; firstly, the newsreader in sound only over a still photograph (the above relevant picture used); secondly, the newsreader heard over the telephone (as though reporting direct from the news location) and the same photograph shown; finally, the newsreader heard over moving film. The choice of subject matter—a fictitious tornado in Oklahoma—was based on the availability of relevant film and photographic material loaned by Liverpool University's Geography Department; the 'irrelevant' photograph was of an indecipherable abstract design. Each of the six presentation conditions, possibly bearing different connotations regarding news immediacy and the efficiency of its reporters, may be suspected to exert differential effects upon viewing attitudes measurable in the normal manner; and certain specific comparisons are possible in this connection. Each version was shown to at least ten student subjects (N = 62 in all); and the newsreader was rated on 13 bipolar scales.

Results

One way analyses of variance conducted as usual on each scale yielded significant effects of presentation on two of them, including the RELIABLE/UNRELIABLE scale significantly affected in the similar Experiment 2 (see Table 3.6). Multiple comparison tests revealed a number of significant differences between conditions on these scales, though of course only those between conditions differing with respect to one production variable at a time are suitable material for discussion. Suffice it to say, of the exhaustive comparisons conducted between general means, that when presented against the 'relevant' background the newsreader was rated significantly more SCEPTICAL than in all of the other conditions except the 'irrelevant' background ($P < 0.05$ at least).

In the following discussion we concentrate on the results of t-tests conducted between the strictly comparable conditions alone (Table 3.7). Performed for once on all of the scales used they indicate significant differences between the 'plain' and 'relevant' background conditions. Thus the performer was rated less PLEASANT ($P < 0.05$) and BELIEVING ($P < 0.01$) when the relevant background was presented than when, in the 'plain' condition, it was not. Additionally he emerged more INFORMED, STRAIGHTFORWARD and RELIABLE in the 'voice over film' presentation than in the 'voice over still photograph' ($P < 0.05$ in each); however, the significance of the latter result obtains on a one tailed basis only. No significant differences occurred between the 'voice over photograph' and 'telephone' conditions. (The means observed on all scales referred to here—whether or not their differences are highlighted by the analyses of variance—are recorded in Table 3.6.)

Discussion

When a professional newsreader was viewed the keyed insertion of a relevant background reduced his impact.

As expected, far fewer image effects were noted on this occasion than in, for example, the comparable Experiment 2. We may suspect that this is due in part at least to the use of the professional presenter. Insufficient

Table 3.6

ANOVAs/1: News report formats – Experiment 12

Scale	*Mean scores* Plain background	Relevant background	Irrelevant background	V/O film	V/O photo	(V/O photo + 'phone)	*F score*
PLEASANT	2.6	3.5	3.3	2.7	2.8	3.3	0.61
INFORMED	2.4	3.6	3.0	2.2	3.2	2.3	1.62
BELIEVING	2.5	3.7	3.2	2.2	2.6	2.7	2.86*
S'FORWARD	3.1	3.5	3.2	1.7	2.9	2.3	1.73
RELIABLE	2.5	3.1	3.5	2.7	3.8	2.8	3.82†
df							5,56

(NB Lower means denote more positive ratings; V/O = voice over; significance levels: *5%, †1%)

Table 3.7

Significance levels of between condition differences – Experiment 12

Scale	Plain vs. relevant background %	V/O film vs. V/O photo %	
PLEASANT	5		
INFORMED		5	1 tailed criterion only
BELIEVING	1		
S'FORWARD		5	
RELIABLE		5	

(NB Only those differences occuring between strictly comparable conditions are tabulated) controls were available for us to be sure of this in terms of the comparison methods used to this point, though as we have hinted in Experiment 10 different analytic procedures will be called upon at a later stage (Chapter 5) providing a further check.

Certainly the presenter, although unfamiliar to the subjects, displayed a confidence and skill sufficient, in theory, to reduce his perceptual ambiguity for them rather as the ambiguity of a lecturer was reduced by relevant information about her in Experiment 9. The relevant visual detail, by comparison with the plain background similarly, seems to have had a sufficiently compulsive impact—more so than the more ambiguous, irrelevant detail—to cause the viewers' attention to the performance accompanying it to be reduced: yet the latter received significantly more neutral ratings than in the 'plain' condition on the PLEASANT and BELIEVING scales accordingly. (Thus the performance and the background, however closely related, may each represent separate 'channels' of information competing for the viewers' attention.) The compulsiveness of visual background detail, as long as it is relevant, is once again suggested, notwithstanding the problems associated at this relatively early stage of research with the prediction of its precise effects Until the independent variables in a field have each been identified and their various potential effects empirically distinguished the element of unpredictability will prevail. Since it is impossible to predict all effects in advance of studying them closely, we can only attempt initially to build up a mosaic of evidence for them, hopeful that ultimately they may be enumerated and controlled.

The notion that a thoroughly skilled performance will divert viewers' attention from other presentation factors even though the performer may be completely unknown previously, combines with the results of Experiment 11 in forcing us to reappraise our criteria for 'maximum ambiguity' (cf. Scheff, 1973) as applied in the research hitherto. While lack of prior knowledge of a performer and his subject matter—the condition laid down in these experiments generally—may clearly increase his ambiguity for the audience, a competent performance given in a recognisable (i.e. unambiguous) professional manner will obviously reduce it. For a stimulus to have truly 'maximal' ambiguity it must give rise to no decisive impression, whether positive or negative, of any type: it must be totally unmotivated and unmotivating as well as unfamiliar. When, for instance, a TV performance fits this description the potential effects of such factors as production design will be at their greatest. We conclude, as previously, that careful variations of the imagery in a TV production can serve particularly useful enhancing functions when less than totally competent presenters are employed for reasons, perhaps, of their specialist knowledge. But it also now follows that the same variations require careful justification in situations where the performer makes a good impression; for in these they may have adverse effects. '(T)he problem is not so simple that one can say it is always better to illustrate news items with pictures . . . Too many general "picture postcards" (may be worse than) if viewers only had the newscaster to look at' (Findahl and Hoijer, 1977, p.16).

An 'overloading' of the visual detail at the performer's expense may have been the basis for Barrington's finding (1972) that a plain background was more conducive to learning than a simulated laboratory setting. If, in the present experiment, we assume the 'relevant' background presentation to have been that providing the most effective competition with the visual qualities of the performance, the fact that ratings in this condition on the BELIEVING/SCEPTICAL scale were significantly different from those of all others except the irrelevant background certainly becomes understandable. The lack of any significant difference between the relevant and irrelevant background conditions would otherwise be one of the experiment's more surprising features—certainly in relation to the differential effects of visual background upon person perception shown by Canter, West and Wools (1974). As in previous work by Schlater (1970), however, the present subjects were apparently quite unresponsive to the intrinsic differences in meaning between the two backgrounds, or at least not consistently responsive. The greater perceived reliability of the voice over film presentation by comparison with the voice over still photograph is less remarkable, in view of the ample evidence from previous experiments concerning the psychological impact of general visual detail (see also Miller, 1969). It is perhaps surprising only that such results should have been yielded on a one tailed statistical basis only, and that previous still/motion picture comparisons (Beach, 1960; Hazard, 1962-63) should have elicited no significant differences at all. The need for agreement by experimentalists on a standard methodology for ongoing use in this field is urgently indicated.

Experiment 13: Autocue and eye contact

During the recording of the videotape material for the previous experiment, a performance variable came to light considered potentially capable of attitude effects in its own right. In using his teleprompt or 'autocue' device the newsreader had adopted the characteristic, unwavering gaze at the camera lens noted in news broadcasting conventionally. This recalled an effect discussed in Experiment 2: the use of autocue by a TV newsreader on that occasion was considered to have constrained his performance and helped to negate audience attitudes towards him. In order to investigate the possibility of such effects in TV presentation, new material was televised permitting a test of viewers' attitudes to a newsreader using various autocue strategies. Since policies regarding autocue usage are varied in conventional production contexts, evidence as to any effects they may exert on presentation impact is likely to have immediate practical implications.

Several unintended effects commonly betray the use of teleprompting techniques even in professional TV contexts. Small head movements and a side to side motion of the eyes indicate quite unmistakably that a performer is in fact reading from a cue card or similar. If the device is nearly though not perfectly aligned with the camera lens, the performer may also appear troubled by a squint! In this and in various other experiments we have relayed the performer's text to him rolling upwards at speaking rate via an overhead projector onto a wall screen, and thence via a video-camera onto a monitor set; the latter is placed immediately below the camera lens that the performer appears to fixate. As long as the distance between the camera/monitor set and the performer is at least 12 feet, a slight (approximately 9 inch) displacement of his gaze from the lens is not apparent. Sophisticated modern autocue technology overcomes even this displacement by projecting the rolling script onto a one way mirror attached to the lens itself. The performer looks in the exact direction of the lens though he sees his words instead: the camera, on the other hand, viewing from the non-silvered side of the glass, receives an unencumbered image of the performer. Such equipment is of course expensive, and careful use of the system we have described above produces indistinguishable results.

Investigations specifically concerned with the amounts of eye contact given to the camera by televised lecturers, and with audience reaction to them, have been conducted previously by Westley and Mobius (1960) and Connolly (1962). Each compared high and low camera eye contact conditions in a manner virtually identical to that we report here, but with non-significant results. As Coldevin (1976) points out, however, the testing instruments they used were limited, especially with respect to perceived performer characteristics. The present experiment sought to fill this gap in the data accordingly.

Procedure

The same news type text and professional newsreader were used as in Experiment 12, and three experimental conditions were created in which the style of performance was consciously varied while production options were held constant. In the first condition the newsreader was instructed to read from the autocue for the major part of the news item, looking down at his desk script only occasionally. In the second he was asked to look at autocue and desk script to approximately the same extent; and in the third he concentrated mainly on his desk script, occasionally glancing up. In all other respects he was requested to keep the three performances as similar as possible (same speed, emphases, etc.), a task that he accomplished with professional precision. The performances lasted between sixty-three and sixty-five seconds, and the total fixations of the autocue in the three conditions varied with uncanny linearity, being 15 seconds (23 per cent), 35 seconds (54 per cent), and 55 seconds (85 per cent) respectively. The newsreader was presented via the same head and shoulders shot against a plain studio background in each condition. The three presentations were then judged by ten or more student subjects as before (N = 32 in all), using 14 rating scales. As in all of these experiments no subject had seen the performer before.

Table 3.8

ANOVAs/1: Levels of camera-eye contact – Experiment 13

Scale	*Mean scores* 15 secs.	35 secs.	55 secs.	*F score*
PRECISE	3.8	1.8	2.5	4.66*
CAUTIOUS	2.0	2.5	3.5	6.77†
RELAXED	5.7	3.4	3.3	11.84†
DIRECT	4.1	2.1	2.3	5.59†
df				2,29

(NB Lower means denote more positive ratings; significance levels: *5%, †1%)

Table 3.9

Significance levels of between condition differences – Experiment 13

Scale	15 vs. 35 secs. %	15 vs. 55 secs. %	35 vs. 55 secs. %
PRECISE	1		
CAUTIOUS		1	5
RELAXED	5	5	
DIRECT	1	5	

Results

One way analyses of variance on each scale revealed significant effects on four ot them (Table 3.8). Multiple comparison tests (Table 3.9) indicated that, when using the autocue to the minimal extent (i.e. looking mainly at his desk script), the newsreader was considered significantly more CAUTIOUS than in the maximal condition ($P < 0.01$) but also less RELAXED and DIRECT ($P < 0.05$ in each). The intermediate condition in turn gave a significantly more RELAXED ($P < 0.05$), PRECISE and DIRECT ($P < 0.01$ in each) impression than that in which the autocue was used minimally, and a more CAUTIOUS one ($P < 0.05$) than that in which it was used maximally. Perceived caution thus diminished as use of the autocue was increased. Perceived precision, however, was maximal when a compromise was achieved between the excessive and under-usage of the technique.

Discussion

Different styles of autocue usage affected a newsreader's perceived caution, precision, tension and directness.

If an impression of greater precision is a prime production criterion, over-attention to either the desk script or autocue is evidently equally to be avoided. When the performer is a skilled user of teleprompt techniques it certainly appears that he may significantly increase his apparent relaxedness and directness in the process. This effect is as commonly supposed. Yet no benefits were conveyed by the maximal as opposed to intermediate use of autocue in the present situation; and it should be noted that maximal autocue usage was associated with a significant increase in the performer's perceived rashness—a possibly adverse effect unlikely to be predicted at an intuitive level. This result tentatively confirms the earlier suspicion—Experiment 2—of a negative aspect to autocue usage; and it supplements the less yielding research of Westley and Mobius (1960) and Connolly (1962). The experiment as a whole suggests that, in newsreading situations at least, it may be sensible for the reader to divide his attention roughly equally between the camera and the desk script whether or not a cuing device is available in the studio. As noted in Chapter 2, the standard TV reporter is not necessarily assumed by the audience to be an expert in the news matters at hand, and his need to refer to notes will to an extent be expected.

This experiment also raises the paradoxical possibility that a relaxed and polished TV performance may in itself have negative connotations—as, in the present circumstance, of 'rashness'. When a producer and performer carefully contrive a relaxed impression—whether by the use of autocue or by avoiding scripted techniques altogether—it is possible that they unwittingly undermine certain other effects that may be intended. We give detailed attention to this possibility in discussing attitude conflicts in Chapter 5. First, however, we investigate a further possibility regarding autocue usage that was indicated by Experiment 3: namely that such effects may underlie the earlier results relating to the direct/profile variation of camera angle.

Figure 3.1 Direct address: Eyes visible – Experiment 14

Figure 3.2 As 3.1: eyes masked

Table 3.10
ANOVAs/2: Camera angle x visibility of eyes – Experiment 14

Scale (ANGLE) (EYES)	*Mean scores* Direct visible	Direct masked	Profile visible	Profile masked	*F scores* Angle	Eyes	Inter-action
PRECISE	2.8	3.7	2.4	1.9	4.59*	0.15	1.86
EXPERT	3.5	3.8	2.8	2.3	4.85*	0.04	0.64
RELAXED	4.5	3.1	4.3	3.0	0.05	4.27*	0.01
df					1,36	1,36	1,36

(NB Lower means denote more positive ratings; *5% significance level)

Table 3.11
Significance levels of between condition differences – Experiment 14

Scale	Direct vs. Profile (visible) %	Direct vs. Profile (masked) %	Visible vs. Masked (direct) %	Visible vs. Masked (profile) %
PRECISE		5		
EXPERT		5		
RELAXED			5	5

(NB Only those differences occurring between strictly comparable conditions are tabulated)

Experiment 14: Eye contact and camera angle

In the course of Experiments 3 and 13 we have acknowledged that the use of autocue to increase a performer's perceived poise may require a distinct degree of skill. The very act of appearing relaxed and unself-conscious while so artificially constrained by studio technology demands the type of ingenuity and concentration that an actor applies on stage. It is quite feasible that the use of autocue by an unskilled performer may add to his perceived tension, detracting from his overall impact rather than enhancing it. Even the most skilled of TV newsreaders are generally unable to conceal the telltale signs that their eyes are engaged in reading; and since in normal face to face interaction the eyes are commonly far more mobile in any case (Argyle, 1975) their 'behaviour' in the autocue situation is likely to be a major determinant of audience reactions to the performer generally.

In Experiment 3 this possibility was raised specifically in relation to the effects of camera angle, since replicated three times. When perceived in profile formats, it will be recalled, performers have been consistently rated as, for example, more EXPERT than when addressing the camera directly; and this effect has been suggested to derive either from (a) the convention whereby experts are typically televised in interview contexts—i.e. presented in profile—while those who address the camera directly are more usually the (non-expert) newsreaders and linkmen; or from (b) the fact that in the profile condition signs of autocue usage conveyed by a performer's eyes are less apparent. The present experiment was designed to test each of these hypotheses in turn.

Procedure

Two simultaneous TV recordings of a single studio performance (one minute duration) were televised using autocue from direct and profile angles as before (see Figure 2.5). The speaker, a lecturer at Liverpool University, delivered information on the fictitious cross cultural subject of 'Hatu art'; he was seen in each condition in a head and shoulders shot against a neutral studio background. Two copies of each recording were then made and in the process a black rectangle was inserted electronically into one of them, masking the speaker's eyes for the total duration of the item. The manipulation is illustrated in Figures 3.1 and 3.2, presenting 'direct' and 'direct masked' conditions respectively: the remaining two conditions—'profile' and 'profile masked' were strictly comparable with these figures in all respects other than the variation of angle. The masks successfully occluded the eyes at all times. The four videotape versions were each shown to ten student subjects and rated on 14 attitude scales.

Results

Particular interest lay in the analysis of expertise ratings in the four conditions, as this was the scale on which significant differences had consistently occurred before. Two way analyses of variance on each of the 14 scales revealed significant differences in the ratings for apparent relaxedness, expertise and precision (Table 3.10). While perceived expertise and precision were significantly affected by the camera angle factor (i.e. greater in the two profile conditions than in the direct conditions), perceived relaxedness was significantly related to the visibility of the performer's eyes: multiple comparisons of direct and profile conditions in turn (Table 3.11) indicated relaxedness to seem greater in the masked as opposed to unmasked conditions ($P < 0.05$ in each).

Discussion

When varied simultaneously camera angle and eye contact had selective effects on performer ratings.

The earlier findings in relation to direct vs. profile camera angles were once again upheld. Moreover, each of the two explanations tentatively offered in this connection before was found tenable in relation to different judgement scales. The results reaffirm the complex basis of viewers' judgements on separate presentation cues simultaneously: see Experiment 11. They also strengthen our conclusion following Experiment 13 regarding a negative aspect to autocue usage. The performer's perceived 'rashness' on that occasion was at its highest when greater attention was given to the autocue (and thus the camera lens) than to the desk script; and it is now apparent that the greater visibility of a performer's eyes alone may increase his perceived tension in its own right. 'Leakage' of personal tension by non-verbal and facial mannerisms in this manner has been discussed by Ekman and Friesen (1969). If wishing to overcome the message of tension from his eyes, a performer should take the trouble to vary his camera eye contact discreetly even when reading from an autocue; he should acquire the artfulness of an actor in treating the TV camera to the repertoire of temporary looks and glances that he would use in normal 'unscripted' social interaction.

Experiment 15: Performance and focal length

Once the fine details of both performance and production technique capable of influencing audience reactions have been identified, a performer may take steps to cover up the behavioural cues most likely to lead to detrimental effects; and the producer may give subtle assistance by attention to other image cues interacting with them. As we have already seen, the producer may emphasise or distract from performance qualities according to the degree of ambiguity characterising them; and his decision will clearly be an impromptu, intuitive one dependent upon his ability to realise the persuasive potential of particular variables as they become apparent in the course of production. We do not underestimate the complexity of this task; and via the research in general we aim primarily to increase an awareness on the part of TV production staff of the types of effect for which to look out (see Chapter 7).

We also hope, of course, to draw these factors to the wider attention of TV researchers. Previous investigations of TV performance effects in particular have been minimal. Calkins (1971) compared student reactions to TV instructors' styles of dress though reported no significant differences. Similarly, McDaniel (1974) found no difference between viewers' comprehension of different newsreaders. Coldevin (1977, p.91) reinforces the general conclusion that 'a rigid "type cast" formula for presenter selection may have only limited utility' in production design, though he points out that a 'young hip' in casual dress with long hair and beard produced a significant change of attitudes towards educational material among an audience of sixteen and seventeen year olds. It is quite probable that a renewed study of long term performance effects using a comprehensive range of techniques for attitude and behavioural measurement will be more yielding than the earlier ones. While the well documented effects of non-verbal and 'body language' factors in human communication (Argyle, 1969, 1975) may certainly be distorted in the TV situation, they are unlikely to be overriden altogether.

Clearly a major problem at the applied level stems from the complex range of interactions possible between image variables. For the TV producer and researcher the problems of recognising image effects are thus the same. The more they appear to arise from a combination of variables, the less easy they are to predict; and both individuals must have the capacity to specify more than one known variable at a time before the wide range of possible interactive effects may be controlled. Moreover, in seeking to interpret image effects each must bear in mind the situation(s) in which they arose. Quite often the circumstance alone may hint at an effect's origin; and, similarly, sensible practical applications of an effect can only be made when the situation at hand is known to be both appropriate to it and stable. Thus the keying of a background image in order to heighten or diminish performance effects in a news context (see Experiments 2 and 12) may be less appropriate in an interview context, and the use of added visual detail in order to increase the educational power of a presentation for children may be quite inappropriate in one intended for adults.

In the present experiment we investigate a production variable suspected to have exerted unintended effects in Experiment 6. In order to vary the detail regarding a performer's use of notes in Experiment 6, shots taken from a greater focal length than that generally employed were inserted. Failure to replicate the results of Experiment 1 using this technique was possibly due to the inclusion of extraneous influential cues in the long shot conditions. We now check the hypothetical power of camera focal length to influence audience attitudes by varying it in an interview context. However, in view of the increasing evidence of interactions between production and performance variables, provision is also made for a simultaneous variation of performance style in the hope of augmenting the experiment's predictive value.

Procedure

Two one minute performances were given by an actor portraying the administrator of a fictitious hospital. In each he explained to an unseen studio interviewer delays in the construction of new hospital premises; he was seen in semi-profile and (though this was not apparent in the realistic performances he gave) continually referred to an autocue presentation of his text in order to ensure that the words spoken were constant to both performances. In the first he was directed to appear 'friendly' and polite, while in the second 'angry' and indignant. Judicious hesitations and 'word stumbles' were built into the performances in order to create an effective impromptu style.

Two simultaneous recordings were made of each performance, one presenting a close up shot of the actor and the other a mid shot (Figures 3.3 to 3.6). The natural consequence of the mid shot focal length was to bring into view the performer's sitting posture and manual gestures. The two cameras were placed immediately adjacent to each other so that the difference in camera angle between them was undetectable. The four videotape versions thus created were each presented to at least ten student subjects (N = 42 in all), and the performances were assessed on 16 bipolar scales.

Figure 3.3 TV interviewee: friendly/close up – Experiment 15

Figure 3.4 As 3.3: friendly/mid shot

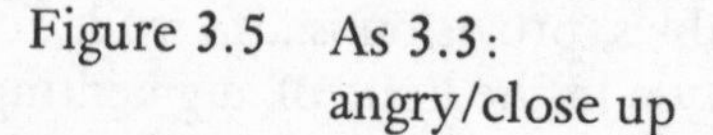

Figure 3.5 As 3.3: angry/close up

Figure 3.6 As 3.3: angry/mid shot

Table 3.12

ANOVAs/2: Performance x focal length – Experiment 15

Scale	*Mean scores*				*F scores*		
(PERF.) (FOCAL)	Friendly close	mid	Angry close	mid	Perf.	Focal	Inter-action
RELAXED	3.1	4.7	4.5	5.4	5.56*	7.01*	0.65
INTERESTING	4.4	5.4	3.8	3.7	6.24*	1.10	1.73
NOT NERVOUS	2.9	4.4	2.6	4.0	0.35	6.44*	0.01
PROFOUND	4.7	5.6	3.9	3.9	7.61†	0.93	0.96
df					1,38	1,38	1,38

(NB Lower means denote more positive ratings; significance levels: *5%, †1%)

Table 3.13

Significance levels of between condition differences – Experiment 15

Scale	Friendly vs. Angry (close) %	(mid) %	Close vs. Mid (friendly) %	(angry) %
RELAXED	5		5	
INTERESTING		1		
NOT NERVOUS				
PROFOUND		1		

(NB Only those differences occurring between strictly comparable conditions are tabulated)

Results

A two way analysis of variance was performed on each scale, and significant effects were found due to performance style and/or focal length on four of them (Table 3.12). These analyses reveal that the speaker appeared significantly more RELAXED to his audience when giving the 'friendly' rather than the 'angry' performance, though less INTERESTING and PROFOUND. When seen in the close up conditions (i.e. regardless of performance style) he was rated significantly less TENSE and NERVOUS than in mid shot. Multiple comparisons of the individual conditions (Table 3.13) disclosed generally predictable differences between performances—but also that in mid shot the apparent tension of the 'friendly' performance was greater than in close up ($P < 0.05$).

Discussion

An increase in camera focal length enhanced a performer's perceived tension even when the style of his performance minimised it.

Presumably the added animation evident in the performer's hands, etc. in the mid shot conditions amplified his apparent tension even when, in the 'friendly' condition, it was minimal. As in the early experiments the results indicate TV's extremely subtle potential as a source of controlled visual suggestion and the conscious manipulation of a performer's impact that may be effected by judicious camera direction.

A direct comparison may be made between this experiment and an earlier one by Wurtzel and Dominick (1971-72). These researchers have also examined the attitudinal effects of televised medium and close up shots, and of an added parallel factor resembling ours of performance style. The two levels on their latter factor related to a 'low key, controlled' style of acting appropriate to TV (cf. our 'friendly' condition) and, by comparison, a more 'expansive and intense' style characteristic of the theatre (cf. 'angry'). Their results—indicating the close up shot to be regarded as more suited to the TV acting style and the medium shot to the more theatrical style—and our own are in essence identical. The criteria for the application of particular shots in various contexts of course must differ: the choice of, for example, a medium shot to emphasise a performer's characterisation in TV drama might well be highly unsuitable in an interview. No significant effects of focal length upon learning have been reported (Aylward, 1960; Cobin and McIntyre, 1961).

As long as they are used responsibly, however, careful variations of focal length may have clear benefits. As Ekman and Friesen (1969) show, movements of the hands and feet and variations in posture betray a person's tension even more plainly than facial expression; thus if the producer wishes to conceal a speaker's evident nervousness before the camera a close up shot may actually be the kindest way of presenting him, and not—as TV producers commonly fear—excessively overfamiliar and intrusive. Of course, extreme close ups upon the face of an interviewee have occasionally been used for notorious dramatic effects, deliberately emphasising the darting eyes and beads of sweat produced under cross-examination. A careful line must be drawn between the beneficial and reprehensible manipulation of visual effects; and the choice of production techniques should be dictated pragmatically according to the nature of the material and performance at hand. The primary moral for those engaged in TV communication is not to discount the highly suggestive impact that small visual aspects of a presentation may have upon those viewing them. The observer, like the TV camera, is highly discriminating and selective, and each should be sensitive to the extraneous impressions formed by the other.

The relationship between higher perceived tension and interest value in this experiment supports an earlier conclusion (see Experiment 13) that the former characteristic is not necessarily all negative! It becomes increasingly apparent that a highly polished TV performance does not always gain the viewers' greatest respect; and a speaker's general credibility in, for instance, instructional contexts may well be enhanced if his apparent tension is emphasised (see Experiment 17). In general we also note an increasing need to determine the relationships between attitude judgements (see Chapters 4 and 5).

Experiment 16: Documentary soundtrack music

A final variable to receive detailed attention in this sequence has been that of musical accompaniment in a documentary film context. Though elsewhere in the book we have emphasised visual presentation aspects, we wish to acknowledge by this experiment that the TV image may involve aural forms of information as well. TV production staff commonly use soundtrack music to create various mood effects, these being the aural equivalent of the context effects we have investigated earlier. Though previous attempts to investigate musical effects upon TV viewers using the semantic differential technique have enjoyed only limited success (cf. Tannenbaum, 1956), a further attempt was made here in view of the greater sophistication of the video presentation technology now available.

Procedure

A three minute extract from a broadcast TV documentary concerning alpine geology was selected, comprising an approximately one minute musical interlude (in which a mountain panorama was scanned) followed by a minute of lecture material by a geological expert, and ending with a further minute's panorama with musical accompaniment. The original pieces of soundtrack music were erased, and four versions of the extract were prepared featuring new musical backgrounds dubbed on as follows:

1. Solemn music (*Marche Slave* by Tchaikovsky) was added to the final segment of the film, the opening one remaining silent.
2. Joyful music (*Marche Joyeuse* by Chabrier) was added to the final segment, the opening one remaining silent.
3. The solemn music was added to the opening segment, the final one remaining silent.
4. The joyful music was added to the opening segment, the final one remaining silent.

In order to determine whether the different musical styles would influence attitudes to the lecture, the four extracts were each presented to ten or more student subjects for rating on 17 bipolar scales. A fifth group of subjects rated a control condition to which no music was added ($N = 52$ in all).

Results

The scores on each scale were inspected via a one way analysis of variance. Barely any difference between the five presentation conditions in general was noted, as in the similar study by Tannenbaum (1956). Only on the RELIABLE/UNRELIABLE scale was a significant difference between conditions evident ($F = 2.90$; $df = 4,47$; $p < 0.05$). Multiple comparisons between conditions on this scale revealed that in condition 3 the lecturer was rated significantly less RELIABLE (mean score 3.9) than in each of the other conditions (mean scores 2.2 to 2.7) at the 5 per cent level at least.

Discussion

The addition of solemn soundtrack music to the beginning of a film documentary extract reduced the subsequent speaker's perceived reliability.

Since this result involves the scores on only one attitude scale, and arises from the atypical rating of just one of the experimental conditions, it may clearly be due to chance alone. The similarly negligible result obtained by Tannenbaum (1956) suggested that musical effects—when they do occur—are highly selective, concentrating upon assessments of such qualities as activity and strength. Thus, as Osgood, Suci and Tannenbaum point out (1957, p.304), they will probably not be evident in results yielded by randomly chosen attitude scales. Osgood et al. go on to suggest that the relationships between attitude scales and the various dimensions of meaning that they collectively represent may yield more reliable predictive information than is ever available when the scores on individual attitude scales are analysed in isolation.

In their own research, Osgood and his colleagues developed extremely influential ideas regarding attitude formation by inspecting their data factor analytically. On the basis of the above analysis alone we naturally do not attempt to draw practical inferences from the data at this stage. Following their re-examination by factor analysis in Chapter 5, however, we point to a further aspect of the data not apparent so far, and also conjecture that the impact of musical context may be more susceptible to fine variations between audience members than that of other variables. Certainly the present statistical methodology, in concentrating on the differences between audiences as a whole, is insensitive to those arising within them; and the factor analysis of these data in due course effectively demonstrates that the difference method by no means exhausts the insights to be yielded by such data. Hence our decision to report this relatively uninformative experiment at this stage.

Further research possibilities

Further possible variables for investigation may be briefly suggested. Work conducted by communication students at Liverpool University on the basis of these experiments has indicated particular effects deriving from, for instance:

1 The sequence in which contrasting news items are presented. In an experiment suggested by the Kuleshov effect reported by Pudovkin (1958) two 'head and shoulders' news reports were recorded, one concerning a flood (serious item) and the other an unusual champion at Crufts dog show (trivial)! They

were edited into serious-trivial and trivial-serious sequences each lasting fifty seconds, with visual jump cuts between them avoided by the use of a picture caption at the beginning of each item. When his two items were viewed in the 'serious-trivial' sequence the newsreader was rated significantly more ABLE and INTELLIGENT by student subjects than in the 'trivial-serious' condition.

2 The convention in TV news requiring reporters seen on film to 'sign off' with their name, programme title and location was examined. A one minute news type report concerning a fictitious archaeological discovery was filmed on a Liverpool demolition site. Two copies of the report were made, identical in all but one respect; whereas one ended with the signature 'Nick Goodwin, News at One, Salford', the other stopped short immediately prior to it and was consequently three seconds shorter. When seen to sign off at the end of his item the reporter was regarded as significantly more MODEST, AWARE, STRAIGHTFORWARD and FRIENDLY.

3 The 'nuts and bolts' approach in the presentation of a studio setting was investigated. Two simultaneous three minute recordings of an interview with a fictitious show business figure were made, one showing the cables and pedestal of a camera apparently in use behind him, and the other framing him slightly differently, as in Experiment 1, so as to exclude these from shot. (While at one time extraneous details of studio activity and equipment were regarded as evidence of shoddy production, the student hypothesised that, as far as viewers at least are concerned, this may no longer be the case.) When seen in the cluttered studio environment the interviewee was rated significantly more INTERESTING and RELAXED.

These demonstrations and the combined practical/theoretical approach to media studies with undergraduate students that they represent are discussed by Baggaley (1978). A fourth variable for possible further study—that of studio seating position in a discussion context—has been suggested by Baggaley (1977). A four minute discussion on 'the state of the nation' was conducted in which one of the participants was physically separated from the other two by a chairman. Two versions of it were televised simultaneously, in one of which the spatial relationships of the participants were clearly disclosed while in the other, via highly selective camera shots, they were concealed. In viewing the isolated participant, subjects evidently responded to the greater detail of his position on the set by rating him on balance less favourably (more WEAK, NERVOUS, UNPERSUASIVE, SHALLOW, etc.). As in Experiment 7, schoolchildren were notably more susceptible to this effect than adults. We are prevented from drawing firm conclusions from this experiment, however, by the results of Experiment 15: in both experiments shots taken from a variable focal length were used, though in the seating position experiment these were not controlled.

The actual range of perceptual variables capable of influencing the TV viewer is of course imponderable. The interactions that may feasibly occur in viewers' responses to the people, words, and other images they perceive on TV are also innumerable. At this stage we might proceed to a study of countless other variables: multiple camera effects, rates of camera cutting, TV colour vs. monochrome, etc.,without substantially increasing our predictive insight into their control for practical purposes. We therefore conclude the present series of experiments satisfied that they indicate:

1 That verbal content, human performance, and visual production characteristics are each capable of individual effects upon audience reaction.
2 That each source of influence may dominate the other two depending upon their relative degrees of ambiguity.
3 That, in attitude terms at least, these effects may be measured and to an extent applied in a predictive practical manner.

We also acknowledge the numerous further ideas for image research of this type offered by Coldevin (1976, 1979).

Ways of increasing the predictive value of attitude data are examined in the next chapter. From an enumeration of the evident and possible types of image variable capable of a psychological impact, we also turn to consider in greater detail the viewer himself. As pointed out by Baggaley and Duck (1976), to overlook the viewers' contribution to TV presentation effects is to ignore the problem; indeed in individual viewers image effects may be defined and determined in entirely different ways, for innumerable psychological reasons that attitude measures alone may be inadequate to monitor. Therefore, if such measures are henceforward to be used in this connection as a reliable predictive index, their relationship to other behavioural measures must be ascertained.

4 Attitudes and behaviour

While attitude data may be both graphic and indicative, it is desirable to show that they relate to other independent measures likely to predict subjects' future behaviour (cf. Aronson and Carlsmith, 1968). Individual attitude scales may clearly be weighted differently in their behavioural relevance for subjects, and until we have compared attitudinal and behavioural data we can have little scope for predicting the psychological importance of image effects in different contexts. How then, in the present research contexts, might such comparisons be made? The informative and/or persuasive TV materials that we have tended to investigate are of the types which normally seek to influence either the viewers' concern for certain issues or their recall for specific facts. Accordingly new measures were designed for the purpose of reflecting each of these behavioural capacities in turn. The extents to which the data they yield predict the attitudinal effects of TV presentation, and vice-versa, are considered in this chapter.

Experiment 17: An appeal to 'charitability'

With the intention of generating a highly diverse set of attitudes for comparison with behavioural data on a common basis, a standard TV appeal format was designed and presented via three highly contrasting styles of performance, and then shown to schoolchildren.

Procedure

Three extracts of approximately one minute each from an appeal were televised, inviting donations to the fictitious 'Virological Research Foundation'. An actor portraying a 'virological expert' addressed the camera directly, while reading an identical script in each condition from an autocue device. He was seen in medium close up (from head to waist), and gave three contrasting styles of performance, one in each condition. In all conditions he occasionally glanced down at a set of notes on his knee in order to counteract the impression that an autocue was in use.

The first version was delivered in a 'stern' and commanding manner, not only in voice but also in gesture. The second was a 'tense' version with a hesitant spoken delivery accompanied by occasional nervous tie fingering and throat clearing. The third was a 'friendly' and confident performance in which the actor spoke neither forcefully nor hesitantly but in the pleasant, straightforward manner of an accomplished TV presenter. The actor practised each style of performance in the studio before recording it until a group of three judges was satisfied that it appeared natural.

The three versions were shown to different groups of schoolchildren, all in the lower sixth form at secondary schools on Merseyside. Between 15 and 17 subjects saw each condition (N = 47 in all). The performer was then rated on 16 bipolar attitude scales. The appeal context used on this occasion also provided an opportunity to compare the attitudinal measures of presentation effect with one of

Table 4.1

ANOVAs/1: Styles of performance – Experiment 17

Scale	*Mean scores* Stern	Tense	Friendly	*F score*
RELAXED	4.6	6.9	4.1	14.45†
HUMANE	3.8	2.3	3.6	4.26*
PLEASANT	4.3	3.4	2.5	4.74*
NOT NERVOUS	4.9	6.8	4.9	11.58†
EXPERT	3.9	6.1	4.5	6.70†
NOT HESITANT	3.3	6.5	4.7	14.21†
STRONG	4.2	5.5	4.0	5.13†
Donations (£)	35.41	37.33	26.13	3.61*
df				2,44

(NB Except for on the donation measure, lower means denote more positive ratings; significance levels: *5%, †1%)

Table 4.2

Significance levels of between condition differences – Experiment 17

Scale	Stern vs. tense %	Stern vs. friendly %	Tense vs. friendly %
RELAXED	1		1
HUMANE	1		
PLEASANT		1	
NOT NERVOUS	1		1
EXPERT	1		1
NOT HESITANT	1		1
STRONG	5		1
Donations (£)			5

'readiness to give money'. Once they had completed the attitude rating task the subjects were presented with a separate sheet bearing the question: 'If you had £100 to give away to charity, how much would you now be prepared to give to the cause referred to in the appeal? *Answer: £.......*'.

Results

A one way analysis of variance was performed on the scores for each attitude scale, and on the amounts of notional money 'donated' in each performance condition. Significant differences were detected on seven of the attitude scales and on the donation measure (Table 4.1). Multiple comparisons (Table 4.2) indicated that the three performances had been correctly interpreted in terms of the intentions behind them. Thus the 'friendly' performance was judged significantly more PLEASANT ($P < 0.01$) than the 'stern' one, and the 'tense' performance appeared more TENSE, NERVOUS, HESITANT and INEXPERT ($P < 0.01$ in each) than either of them, also more WEAK ($P < 0.05$ at least). Less predictable was the 'tense' performance's rating as significantly more HUMANE than the 'stern' one ($P < 0.01$). Nor was there evidence to suggest a general preference for the 'friendly' condition as might have been expected: on all but the pleasantness scale, ratings of the 'friendly' and 'stern' performances were statistically indistinguishable.

In terms of the donation measure the most effective condition proved paradoxically to be the 'tense' one and the least the 'friendly' one: the 'stern' performance was nearly as effective as the 'tense' one though not significantly so.

Discussion

A friendly TV appeal was less effective in behavioural terms than a tense appeal.

As in Chapter 3 evidence is revealed of conflict between reactions simultaneously formed by viewers, this time in relation to variations in TV performance style. That the style considered to be the most PLEASANT was significantly less effective than the 'tense' one on a (quasi-behavioural) money-giving index might appear bewildering. Yet it appears that subjects associated the 'tense' style of performance with a high degree of humanity. In a professional TV presenter evident tension, however qualified, is certainly unlikely to increase viewers' belief in his suitability for the job; in an expert from a different professional field, however, evident humanity is far more likely to be a persuasive criterion than a high degree of confidence and competence in the TV studio. Indeed in a serious matter such as a medical appeal an easy, relaxed performance may actually detract from a presenter's image of professional integrity, whether he is presumed to be a competent TV performer or not. On this basis the superior efficiency of the 'tense' performance in behavioural terms is certainly more predictable than at first glance. The evident associations between perceived tension, sincerity, and straightforwardness reported in Experiment 5a point to the same conclusion.

The results thus indicate that the selection of TV presenters simply according to their professional poise before the camera may be totally inappropriate in certain contexts. Different production situations generate quite different expectations on the viewers' part, and in attempts by producers to fulfil them different criteria need to be observed. The full extent of situations in which a friendly and confident performance may have negative associations as above remains to be seen. The most extreme hypothesis for testing in this connection would be that all situations in which a performer addresses the TV camera in an apparently relaxed and natural manner may be open to suspicion; the assumption that viewers disregard the essentially contrived nature of this type of performance in forming their judgements of it certainly requires examination before such an hypothesis may be overruled. Future research may therefore usefully seek to identify the factors underlying attitude judgements, and the possibly conflicting relationships between them (see Chapter 5); and the psychological importance of attitude scales in various contexts should be investigated in relation to independent measurement criteria as here.

Experiment 18: The donation measure—BVT (students)

It has been possible to test the predictive powers of the 'donation' measure of presentation impact in three of the other experimental contexts reported earlier. Experiments 6, 7 and 8 were in fact the only other studies in this series in which an appeal format appropriate to this type of measurement has been used. Donation data yielded in response to the 'British Vagrancy Trust' appeals during Experiment 6 are now reported (see also Baggaley and Duck, 1979).

Procedure

When the subjects in this experiment—60 undergraduates at the University of Lancaster—had completed their attitude assessments of each 'BVT' appeal they were posed the question regarding their 'willingness to give money' first used in Experiment 17. Their responses were obtained on the range £0–100 as before.

Results

(NB Throughout this chapter the variance ratio F is notated as F' in order to distinguish it from condition F in the 'BVT' experiments.) The sums of notional money given in the six presentation conditions were subjected to a two way analysis of variance as in the treatment of the attitudinal data previously. A significant main effect was noted, due primarily to the significant differences between levels of 'visual detail' as expressed by the three columns of Table 2.1. The effects due to 'format' and interaction factors were not significant. Multiple comparisons between conditions revealed the major source of these effects to be condition E (Direct + Notes + Location), the recipient of the highest average donation: £22. This condition was significantly more efficient ($P < 0.01$) than D (Profile + Reaction), the least efficient money raiser receiving an average of £4.40! Average donations to A, B and C were between £7.20 and £8.50, and to F £13.10. Of these E was significantly more efficient than A, B and C ($P < 0.05$ at least).

These results may be compared with the attitudinal findings already reported in Experiment 6. It may be recalled from Tables 2.2 and 2.3 that, on all four of the attitude scales significantly differentiated by visual presentation in that experiment, the most favourable ratings of the performer were given in condition F (Profile + Reaction + Location). On all but the 'strength' scale the condition associated with the least favourable attitude ratings—significantly so by comparison with F– was C (Direct + Notes). Clearly the attitude measures used do not predict the behavioural measure at a simple level, nor vice versa; for the presentation conditions associated with the most extreme ratings on the significantly differentiated attitude scales do not coincide with those receiving the extreme ratings on the donation measure. Yet the results on the two types of measure are not as incompatible as they appear. Via the attitude measurements in Chapter 2 we have already determined the separate (previously controlled) presentation variables actively influencing viewers' reactions, and on this basis a predictive relationship between the two measures becomes apparent, as the following discussion indicates.

Discussion

Ratings on the donation measure by students were at their most positive when the significant effects of image cues upon their attitudes summated.

For the student audience of the 'BVT' appeal–Experiment 6–two of the image variables manipulated have been shown to exert significant attitudinal effects: i.e. the location shots and the long shots including the presenter's use of notes. The addition of these two types of detail evidently enhanced ratings on the strength and straightforwardness scales respectively. The only presentation condition containing both types of cue was E (Direct + Notes + Location); and we conclude that the subjects awarded E the highest amount of notional money on this basis. It is suggested that their behavioural judgement in this condition was influenced by a summation of positive visual cues, and may be predicted by ratings on the 'strength' and 'straightforwardness' scales already shown to differentiate this particular condition from others for this audience (Tables 2.2 and 2.3).

This notion is consistent with the 'cue summation' theory applied by Severin (1967). The theory's basic premise is that the amount learned from a presentation increases in relation to the number of meaningful cues within it. If, as by attitude ratings, individual cues can be shown to contain appreciable meaning for the subjects, we are in a position to predict that a condition combining complementary cues will be more influential on an independent basis than conditions constructed more arbitrarily. The present experiment suggests this to be the case not merely in basic learning terms but also with regard to the amounts of credence placed on what is learned and to the subjects' likely future behaviour.

A complementary, or summative, effect by image cues may in theory be either positive or negative; and if as in condition E of Experiment 18 two positive cues combine (Notes + and Location +) a summative effect of a positive type may be predicted on the independent measure (viz. the £22 awarded in notional donations). If, on the other hand, negative cues combine–or as in conditions A, B and D the positive cues are lacking–a relatively adverse effect may be predicted on the same basis (viz. the significantly lower donations of £8.50 and less). When only one positive cue is in evidence as in F (Location +, but no Notes) an intermediate effect may be expected: hence the £13.10 (a sum which proves strictly moderate, since neither higher nor lower than others to the criteria for a significant difference).

In relation to condition E's behavioural superiority over F, the cue summation rationale thus proves a more reliable predictor than the basic comparisons conducted on the attitude scales individually in Experiment 6. It also proves effective in explaining the behavioural inferiority of condition D (no meaningful cues) in comparison with C (Notes +). For C, though the least favourite condition according to 3/4 of the earlier attitudinal criteria, has the positive effect of the 'notes' cue on the strength scale in its favour, if on no other. The influence of 'notes long shots' in the experiment generally was defined, indeed, by its enhancing effect in this specific connection.

For the media producer the basic predictions of cue summation theory with regard to image design are in some respects less than astonishing. They indicate in brief that better production leads to better effects! In other respects, however, they lead to observations that intuitively may be far less apparent. Having already deduced that individual image variables such as 'notes long shots' may have a selective impact upon one particular scale from a repertoire of many, we may generate the following corollaries: that, when different image cues summate, the attitude scales known to be affected by them will have, firstly, a greater psychological importance for subjects than other scales, and secondly, greater importance than in more arbitrary conditions featuring fewer or none of these cues. A successful test of these two corollaries at this stage would have several benefits. It would provide a basis for distinguishing the scales likely to be most useful in future research from others with lesser psychological importance; and it would yield criteria for investigating the selective effects of the TV image on different types of viewing response.

Experiment 19: Cue summation and attitude weighting

The effects of cue summation as posed so far are, as it were, higher order behavioural effects predicted by the combined impact of individual image cues at possibly very selective lower levels. This rationale generates the two corollaries given above which are now tested via the data of Experiments 6 and 18. Did the 'strength' and 'straightforwardness' scales—significantly affected by two of the image variables manipulated in Experiment 6—indeed differ in their subjective weighting from the other scales, and if so do their weightings in different conditions vary systematically with respect to the behavioural rating scale used in Experiment 18? Certainly the general importance of 'strength' as a factor influencing attitude judgements has been demonstrated since the pioneering work on the semantic differential technique by Osgood, Suci and Tannenbaum (1957). In a highly formative contribution to attitude research they indicated that attitude judgements are commonly related on a series of separate structural dimensions; and in the attempt to pinpoint the factors determining subjects' attitudinal and behavioural tendencies in TV situations, the need to examine the dimensions underlying our own semantic differential data—both between and within the scores of individual subjects—now becomes crucial.

Via factor analytic methods, Osgood et al. established (see Chapter 1) that the variance within a set of data is commonly accounted for by three main factors, relating to a broad 'evaluative' assessment of the stimulus and to further judgements of its 'potency' (i.e. strength) and 'activity'. The experiments in our own series have typically featured scales thought to represent each of these main factors (cf. Baggaley and Duck, 1976, p.85); the relative importance of the factors, however, is known to vary in different situations, and individual stimuli may be variously judged in terms of one or more of them (Warr and Knapper, 1968). Studies of communicator influence by, for example, Tannenbaum (1955), Andersen (1961), Berlo (1961) and Williams (1963) have indicated selective effects upon evaluative and dynamism judgements; though as Markham's results indicate (1968, p.62) even the adjective scales most central to such interpretations (e.g. GOOD/BAD, STRONG/WEAK) can gain radically different meanings, by association with other scales, in different contexts. The more complicated or ambiguous a stimulus is for those assessing it, the less it is likely that their judgements are represented by a simple E–P–A type of dimensionality at all; and in order to test whether, for the student viewers of our 'BVT' presentations at least, particular weight of meaning is attached to the perceived strength and straightforwardness of the communicator's performance as predicted, the attitude ratings yielded in Experiment 6 were now factor analysed.

Procedure

The ratings obtained in each of the six 'British Vagrancy Trust' conditions in Experiment 6 were intercorrelated by the product moment technique (Bruning and Kintz, 1968, section 4.1) and subjected to a principal factoring technique involving an iterative procedure as discussed by Nie, Hull et al. (1975, section 24.2.2). Notwithstanding the sophisticated computational procedures nowadays available for factor analysis, its use invariably demands several subjective decisions. Iteration, for example, is a procedure recommended when the number of variables analysed is relatively small (Comrey, 1973, pp.73-5) as in the case of the attitude scales investigated here. A further decision relates to the number of factors to be extracted from the data. In theory, computations may be continued until the total amount of variance in the data has been accounted for, yet beyond a certain point successive factors may have little psychological meaning, and a criterion must be adopted dictating the point at which useful factoring is to cease. In the factor analyses reported throughout the following sections the conventional 'eigenvalue-one' criterion (Kaiser, 1960) has been applied. In each analysis any one factor consequently represents a pattern of interrelationships which in psychometric terms may be considered significant. The factor initially extracted from a set of data represents the largest pattern of relationships underlying it (i.e. that accounting for the highest proportion of variance in the data); the second represents the next largest unrelated pattern: and so on (Rummel, 1967). The concept of the 'eigenvalue' or latent root of a data matrix is discussed in full by Rummel (1970, pp.95-100).

When the initial factors have been extracted and the most comprehensive patterns of variance in the data are thus decided upon, the most prominent clusters of interrelated data may be highlighted by a rotation procedure. The principles of factor rotation are discussed by Rummel (1970, Chapters 16-17). In the present chapter we assume, with Osgood et al. (1957), that the factors underlying our data are independent of one another (i.e. orthogonal) and the Varimax method of rotation (Kaiser, 1956) is used accordingly. The degree of relationship between a factor and each individual attitude variable is indicated by the scale's factor 'loading', a coefficient ranging from +1 (maximum relationship in a particular direction) to -1 (maximum relationship in the contrary direction). It should be stressed that the loadings on a given factor bearing a minus sign do not necessarily reflect the negatively oriented ratings in the raw data. They indicate that the scores on a particular variable have an inverse relationship with those on variables bearing the plus sign; and in the present context oppositions of this type indicate conflict between ratings tending towards the lower extreme of a scale (denoting the adjectival quality presumed positive) and ratings on other scales tending towards the higher extreme (and the qualities presumed negative). Which of the scales is rated positively and which negatively, however, the analysis does not specify (see Chapter 6).

The criterion for deciding whether a factor loading is high enough to denote a meaningful factor/scale relationship is once again somewhat subjective; the factor analytic interpretations of the present data are based on loadings of plus or minus 0.55 or above, denoting a 'good' or greater factor/scale relationship (Comrey, 1973, p.226). In an orthogonal solution the square of a given loading multiplied by 100 indicates the percentage variance on a particular scale for which a given factor accounts, and the percentage of total variance in the data that is systematically accounted for by the factor as a whole is equal to its eigenvalue x 100 and divided by the total number of scales used (Rummel, 1967, p.466); unless the extraction of initial factors is completed prior to iteration the eigenvalue of a factor is in turn equal to the sum of squared loadings upon it (Nie, Hull et al., 1975, pp.477-8).

Since in the present context we choose to overlook those factors falling beyond a specified cut off point, the variance percentages quoted for each factor in the discussions that follow relate to their share of the common variance: i.e. that accounted for solely by the factors retained. In our own data this usually comprises in the order of 90 per cent of the total variance in the data (see Appendix 4 obtainable from the author). The percentage common variance for each factor is estimated as its eigenvalue x 100 and divided by the sum of the eigenvalues for the whole factor matrix. This and the preceding estimates of percentage variance are precise for all orthogonal factors and approximately so, where relevant, for correlated or 'oblique' factors (Rummel (1967, section 4.3).

The unrotated factor loadings derived by the principal factoring of viewers' reactions to the six conditions of Experiment 6 were next subjected to an additional computation designed to test the cue summation corollaries of Experiment 18. As a first step towards determining whether the scales measuring perceived strength and straightforwardness (the 'cued' scales in this instance) carry, in given conditions, greater than average psychological importance for the subjects, we have compared the mean variance of these scales on each factor with that to be expected for each scale if the common variance accounted for by the factor as a whole were equally shared between all scales. The resulting formula is essentially an F-test for the homogeneity of variance (Edwards, 1965, Chapter 8) in which

$$F^{\check{}} = \frac{\text{mean variance (cued scales on each factor)}}{\text{mean variance (the factor as a whole)}}$$

$$= \frac{\alpha^2}{n'} \div \frac{\text{eigenvalue}}{n}$$

where n = total number of scales; n' = number of cued scales; α^2 = sum of (Σ)

common variance on the cued scales (= Σ squared loadings on the cued scales); and the common variance on the factor as a whole is represented by its eigenvalue. Simplified, the formula becomes

$$F' = \left[\frac{\alpha^2}{n'} \times \frac{n}{\text{eigenvalue}} \right] = \left[\frac{n\alpha^2}{n' \text{ eigenvalue}} \right]$$

In each of the six conditions the values computed via this ratio for the series of separate unrotated factors were next summed and averaged to produce a quotient comparing the observed and expected variances of the cued scales for the full factor solution. Thus derived the quotient expresses the relative weighting (RW) associated with the cued scales in the condition (x) as a whole. Where

N = number of factors,

α_i^2 = the value of α^2 for factor i,

eigenvalue_i = the eigenvalue of factor i, and

$\sum_{i=1}^{N}$ = the total comparison for the N factors, beginning with factor 1, where i = 1,

the full equation is expressed as

$$RW_x = \frac{\sum_{i=1}^{N} \left[\frac{n\alpha_i^2}{n' \text{ eigenvalue}} \right]}{N}$$

Unrotated factor loadings are examined in this context since in seeking to calculate the relative weightings of particular attitude scales—and indeed in inspections of our results that follow—we are concerned less with the relationships within particular attitude clusters than with the importance of individual scales in the data as a whole. Both the order and magnitude of rotated factors and the size of their loadings may be quite independent of the comprehensive data patterns (Rummel, 1967, pp.466-7), and the rotated solutions are therefore inappropriate indices in the present context.

Results

General results of the six factor analyses are reported first. In each of the six presentation conditions, the students' judgements were found to be based on at least five significant factors (see Appendix 4, obtainable from the author). In all conditions, factors easily identifiable with the straightforwardness scale were noted, and in all except D—least favoured on the donation measure— with perceived strength also. Indeed, in C—the least favourable condition in attitudinal terms—these two were the only scales that were capable of clear identification with specific factors in either the unrotated or rotated solution. They dominate the fourth and fifth unrotated factors in this condition respectively.

In condition D, on the other hand, it is observed that over 50 per cent of the variance underlying the subjects' judgements—i.e. more than that accounted for by one factor in any other condition—was related to an (unrotated) first factor identified to a marked degree with judgements of STRAIGHTFORWARD-CONFUSING. No factor in this condition was clearly identified with perceived strength. Substantial loadings by 12/16 attitude scales on the first factor, and the consistency of their directional signs, indicate a classical evaluative dimension: viz. a tendency towards general agreements in the usage of different scales both by individual subjects and in the subject group as a whole. The second factor in condition D, however, is less harmonious than this, pointing to a tendency for positive ratings of the performer's humanity to be associated with negative ratings of his persuasiveness (and vice versa).

The judgements in condition E–most efficient as a notional money raiser–yield a similar if less prominent first factor once again identified with, among other **scales**, straightforwardness. Notable second and third factors in this condition each account for over 20 per cent of the common variance; they are primarily associated with perceived pleasantness and persuasiveness respectively. Judgements of pleasantness on the second factor were inversely related to those of strength.

In F–the condition rated most favourably in attitudinal terms–the straightforwardness scale is once again prominent on the first factor; however, by comparison with the corresponding factor in E, this factor's percentage share of the common variance in relation to that of the other factors is markedly low. Alone of all other first factors, it is also characterised by a significant inverse relationship between straightforwardness and, for instance, perceived profundity: also, most notably, by one between straightforwardness and strength.

The predicted psychological prominence of the latter two scales in the subjects' responses generally is thus convincingly indicated, though more strikingly in certain conditions than in others. When the unrotated factor loadings relating to straightforwardness and strength ratings were subjected to the relative weighting analysis, the weighting quotients obtained–one for each of the six conditions–were found to differ systematically with respect to the average amounts of notional money donated to them. The computed values of RW and the average donations for each condition are presented in Table 4.3. A significant linear relationship is observed between them (Pearson's $r = +0.94$; $df = 4$; $P < 0.01$). (An RW value greater than unity denotes a greater than average weighting for the cued scales within the factor solution generally; while $RW < 1$ indicates a lesser than average weighting.)

Table 4.3

Relative weighting quotients (cued scales) and donation means – Experiment 19

	'BVT' conditions					
	A	B	C	D	E	F
RWs (STRONG + S'FORWARD	0.908	0.832	0.711	0.671	1.145	0.982
Donations (£)	8.50	8.20	7.20	4.40	22.00	13.10

Discussion

The students' attitude ratings significantly affected by specific image cues had a greater than average psychological weighting exclusive to the condition in which these cues summated.

Inspection of Table 4.3 indicates that both of the cue summation corollaries indicated by Experiment 18 have been substantiated, in the students' data at least. In the one condition (E) containing both of the image cues significantly enhancing these subjects' reactions (notes long shots and location detail) the two cued scales, strength and straightforwardness, had, first, a greater importance for the subjects than the average expected for the full repertoire of scales, and, second, greater importance than in all other conditions, including F containing the positive 'location detail' cue only. Furthermore, the average psychological importance of the two scales in the six conditions generally is seen to decrease on a par with the mean scores yielded by the donation measure. The variations in visual imagery between the conditions have apparently exerted differential effects on selected attitude scales; and, when complementary to one another, these have summated to presage a higher order behavioural outcome. The higher the relative weighting of the cued scales in the experiment, the more consistent and thus predictive their scores are considered in this respect.

In certain situations—or for certain subjects—individual variations in visual technique may naturally be quite inadequate as cues to behavioural effects of any significance whatever; though when judiciously combined their impact may clearly be substantial. It is also clear that the earlier between condition comparisons of scores on individual attitude scales, while valuable as initial indices of the image cues exerting significant lower level effects, overlook a wide range of interscalar relationships upon which higher order effects may be based. In disclosing a predictive relationship between a quasi-behavioural measure of response and the structional dimensionality of the subjects' attitude judgements, the present results illustrate that, via factor analysis, the meanings and importance of individual attitude scales may be highlighted and effects revealed that straight polling type questions or attitude scales analysed in isolation might completely disregard (cf. Osgood et al., 1957, pp.309-10). They also recall the findings by Tannenbaum (1955) concerning the particular role of the 'potency' dimension in audience reactions to the TV coverage of a congressional hearing, and by Mandell and Shaw (1973) in relation to the 'potency' effects of a low/high camera angle.

Further computations of relative weighting for each individual attitude scale, and their comparison with the behavioural data via, for example, multiple regression analysis (Nie, Hull et al., 1975, Chapter 20), would certainly yield much additional information regarding the situation-specific importance of attitude measures in general. In condition F, for instance (Profile + Reaction + Location), we have evidence for a relatively high weighting in the case of the straightforwardness scale, which, as the DF effect in Table 2.3 has shown, was specifically associated with the addition of the 'location' detail to condition D. By gauging the subjective importance of an individual scale in each condition in turn, the relative weighting formula can determine the primary origin of a differential effect such as this precisely. In condition E the straightforwardness scale is evidently less prominent. Yet E is still logically more capable of the suggested summative effect than F; for the ratings of strength and straightforwardness in E are at very least independent, while in F—despite its general favourability according to the analytic criteria of Chapter 2—they directly conflict. Unrelated scores are, not surprisingly, presumed to be more capable of a summative behavioural effect than conflicting ones.

As on several earlier occasions, therefore, the centrality of attitude conflicts in our data is indicated—also the need to cater for them as carefully in analysis as the more harmonious patterns. It is worth noting that if the rotated factor solution had been emphasised in the above analysis the evidence of conflict in the general pattern would have been distorted if not concealed altogether. Yet the common practice by social scientists using factor analytic methods is to stress the 'terminal' rotated solution, regarding the unrotated ('initial') matrix as a mere step in the process, and paying little heed to its implications. In this they are encouraged by the standard source books in the field:

> While the principal factor solution . . . is unique in the mathematical sense, at least to psychologists it is not acceptable as the final form (Harman, 1967, p.93).
>
> Although (it) gives a factor solution based on mathematically acceptable factor constructs, the factor constructs represented in an unrotated factor matrix are rarely useful in scientific work (Comrey, 1973, p.9).
>
> (T)he extracted factor matrix does not usually provide a suitable final solution to a factor analysis problem (ibid., p.103).

But when Thurstone (1947) pointed to the usefulness of factor rotation in attempts to highlight the data patterns of greatest 'psychological meaningfulness', it is unlikely that he intended it to be used henceforward at the risk of obscuring other phenomena merely less amenable to immediate interpretation. It is unfortunate for psychological research in general that the impulse to classify human processes at relatively simple levels has, in the years since, led this to be the case.

When the discrete attitude clusters between which tensions and conflict may occur are the prime objects of analysis as in Experiment 24, factor rotation is a valid and necessary manoeuvre. For the present, however, the purpose of our analysis is to focus attention upon audience phenomena at a broader level, embracing the more peripheral as well as general processes among them so that practical steps may ultimately be taken to maximise TV's benefits on as comprehensive a basis as possible. Until its various side effects have been defined any communication medium must naturally have a wide range of arbitrary and ambiguous effects, which will be reflected in the inconsistency and ambivalence of audience responses. In the next section, therefore, we consider the question of image ambiguity in greater detail.

Experiment 20: Arbitrariness vs. motive

It is appropriate here that we should recall the notion of maximum ambiguity (see Experiment 12). When a stimulus has this quality we have defined it as essentially giving rise to no decisive impression, whether positive or negative, of any type: not only must it be unfamiliar, but also totally unmotivated and unmotivating. In comparing attitude scales above as to their degrees of psychological weighting and their association with particular image cues, we are in effect describing the extent to which image variations have affected the ambiguity of a presentation, either by diminishing it (as by cue summation) or maximising it (as in the conditions when certain meaningful cues are lacking). The question is closely related to that discussed by the semiologist de Saussure (1915) in his classification of designated meaning from 'the arbitrary' to 'the motivated'. Baggaley and Duck (1976) have suggested that measurement techniques should be developed in media research for probing these basic semiological ideas further,and in particular 'the probability of particular interpretations of a (presentation) code under specified conditions' (p.156):

> . . . control over communication consists of maximising (by training procedures, for instance) the probability that the intentions conveyed by a given code are interpreted correctly, and it is from study of the contexts in which codes are maximally successful that the ability to predict and control communication effects will derive.

The previous experiment was, consciously, a first step in this direction. In relating two of the image cues present in the 'British Vagrancy Trust' presentations to a series of significantly enhanced attitudinal and behavioural effects, we have identified them as having 'coded meaning' for the subjects of Experiment 6 on specific measurement scales. The condition (E) in which these cues summated was shown to be minimally ambiguous (or completely 'unarbitrary') in this experimental context, and to 'motivate' particular reactions.

On the same basis we may now reconsider the two conditions yielding the least positive attitudinal and behavioural scores in Experiment 6: C and D respectively. The overall inferiority of condition D on the donation measure was accompanied by the absence of any factor identifiable with perceived strength–the attribute evidently cued by 'notes long shots'. However, C, as indeed all other conditions except D, yielded a factor pattern within which the strength scale loads highly. We may attribute its marginal superiority over D on the donation measure to this fact. Just as condition E was found behaviourally superior to F owing, in theory, to the unusually high weighting of the cued reactions to it, in D a totally negative behavioural effect is identified with minimal response consistency–and thus maximum ambiguity—on the same two cued scales (Table 4.3). The analysis of relative weightings in the various conditions thus gives direct information as to the degrees of arbitrariness/motive, as measured by particular scales, in each one.

The concept of arbitrariness has already become central to our issue via the corollaries derived in Experiment (18). As de Saussure himself indicated, however, no image—visual, verbal or whatever—is ever likely to be wholly arbitrary. It was recognised in Chapter 1, indeed, that the individual faced with increasing ambiguity is likely to base new, often unsubstantiated inferences on cues to meaning that he would otherwise have regarded as quite superficial (Festinger, 1954).In order to gain the fullest understanding of the decisiveness of image effects in different conditions, therefore, it is necessary for us to examine the relative weightings not only of those attitudes known to be significantly affected by identical image cues, but also of all attitudes found by the between condition comparisons to have varied in response to image manipulations generally, whether or not the precise source of such effects is known. In the original 'BVT' experiment, for example Experiment 6, we have observed significant between condition differences not only on the two cued scales (strength and straightforwardness) but also on the relaxedness and expertise scales. Although we have attributed each of the latter to 'an unspecified interaction of cues' we must nonetheless embrace them now.

In many practical situations the problems of identifying the precise origins of an attitude effect may indeed be more pervasive than we have encountered so far in the experimental situation. For the theoretical purpose of Experiment 19 it is fortunate that a series of presentations containing tightly controlled experimental variations was available, making specific identifications possible. But, as a check on the applicability of the basic RW formula in situations where this is not necessarily so, we now conduct a further analysis in which the average relative weighting per condition of all scales known to have been differentiated by presentation in Experiment 6 is computed, and in which the resulting RW values are then compared with the associated donation means as in Experiment 19. We make no prediction of the outcome of this analysis and merely seek to establish whether, following the new procedure, a systematic relationship between the attitude weightings and behavioural scores is still in evidence.

Procedure

The precise between condition effects noted on the four attitude scales significantly influenced by presentation strategies in Experiment 6 were re-examined. Reference to Table 2.3 shows that on the strength and straightforwardness scales all 'BVT' conditions were involved in significant between condition effects, for reasons sometimes if not always empirically obvious. On the expertise scale differences occurred between each condition and at least one other, with the sole exception of A (Direct).

The fourth scale affected by image variations—that measuring perceived relaxedness—discriminated between reactions to B and F, and to C and F.

In the case of condition A, therefore, only the strength and straightforwardness scales can be considered to have been influenced by presentation variables: the RW for all attitudes significantly differentiated in comparisons involving this condition will therefore remain as computed in Experiment 19. Reactions to D and E, however, are distinguished from those given to other conditions on three scales, and the RW values for D and E must be recomputed accordingly: likewise in the cases of B, C and F, each of which is differentiated from at least one other condition on all four of the scales.

Results

In each of the six conditions the average weightings of all attitude scales differentiating the condition from at least one other were compared with their expected weightings by the RW formula given in Experiment 19. The RW values obtained for each condition were then plotted graphically in relation to the associated donation means (Figure 4.1). While a systematic linear relationship between relative weightings and donations is observed for conditions C, B, A, F and E—as in Experiment 19—the remaining condition D (least favourite on the behavioural measure) deviates from this pattern. In fact the relationship of all six conditions is more V-shaped than linear, and may be interpreted accordingly.

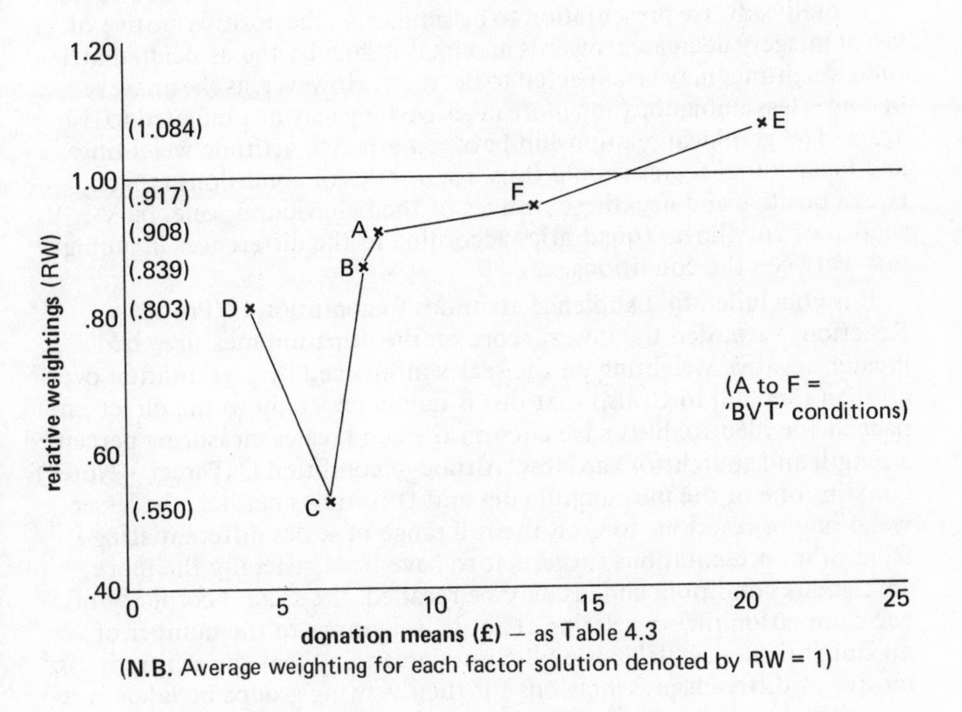

Figure 4.1 RW quotients (all differentiated scales) vs. donation means – Experiment 20

Discussion

The average weightings of student attitude ratings significantly differentiated by presentation variables in general approximated to a curvilinear trend interpreted in terms of the relative ambiguity of the stimulus materials rated.

As noted in the initial discussion of cue summation theory in Experiment 18, image effects range ultimately from extreme positive to negative (favourable to unfavourable). Since none of the physical cues inserted into the 'BVT' material had totally negative effects upon the audience reactions so far examined in the light of this theory, we have not been able to illustrate the predictable effects of negative cue summation in terms of our data. If a totally adverse image effect were created, however, we would naturally expect it to be as decisively unambiguous as a totally positive or favourable one. Since the weighting of the attitude scales visually cued in condition E—regarded by the subjects of Experiment 18 as unambiguously positive—is found greater than in the other more arbitrary conditions, we may expect the RWs of the scales affected by a totally adverse presentation to be similar. As the positive motive of visual imagery decreases towards maximal ambiguity the associated attitude weightings may be expected to decrease. However, as the imagery becomes less ambiguous and more negative they may be predicted to rise again. The graphical relationship between affected attitude weightings and behavioural scores should thus, for a series of conditions ranged between positive and negative extremes of the behavioural scale, be V-shaped or curvilinear (quadratic) according to the differences in ambiguity between the conditions.

It is concluded that audience attitudes to condition D (Profile + Reaction)—awarded the lowest score on the donation measure—had a greater negative weighting on the scales influenced by presentation overall than they did to C; also that this is due in part only to the direct impact of the identifiable image cues on the cued scales measuring perceived strength and straightforwardness. Although condition C (Direct + Notes) contains one of the meaningful cues and D contains neither, the lesser weighting of reactions to C on the full range of scales differentiating it from other presentations suggests it to have been generally the more ambiguous condition; and, as may be recalled, the second corollary of cue summation theory relating attitude weightings to the number of meaningful cues available is explicit in demanding that the arbitrariness/motive of different presentations for their viewing groups be taken into account when higher order effects are predicted from lower order data. (Thus, when particular image cues summate, the attitude scales known to be affected by them will have greater importance than in more arbitrary conditions containing fewer cues or none at all.) The greater relative weighting of all differentiated attitudes to D by comparison with C is therefore argued to be consistent with a systematic relationship between attitudinal and behavioural image effects of the curvilinear type described above.

Ambivalence vs. indifference

The prediction via attitude measurements of higher level image effects thus requires that the subjective importance of individual attitude scales in the relevant condition be estimated, and also that the ways in which individual image cues systematically combine to influence them be ordered theoretically. Factor analysis yields indices for both of these purposes, suggesting a quadratic relationship between attitude weightings and behavioural scores which has been argued to represent the underlying system in question. This in turn is essentially a quantified expression of the relationship between signs and meaning debated by de Saussure (1915) in terms of 'l'arbitraire'.

One difference between the classical notion of arbitrariness and our own, however, is as follows. The classical semiological paradigm involves a fixed idea–the signified–and its communication in symbolic form via a signifier. In semiological parlance the arbitrariness of a sign represents the degree to which it logically reflects the idea signified: if the sign is fully motivated, arbitrariness is minimal and the signified is expressed analogically (e.g. in pictorial form). Yet in the present context the arbitrariness concept proves equally germane to the investigation of ideas formed by visual imagery whether they may be traced back to an actual signified or not. Moreover, the present usage assumes neither that the ideas conveyed by visual cues are deliberately intended by those who design the TV image, nor that they are consciously apprehended by those who interpret it. It is nonetheless supposed that, if sufficient cues gain complementary meanings in parallel, they may summate to effect psychological consequences as significant as any derived via the more obvious verbal means of communication.

The theory of cue summation thus rationalises the processes operating when image cues have subjective meanings complementary to one another. We have considered the systematic way in which image variations can maximise visual ambiguity by failing to provide meaningful cues, and conversely the way in which the summation of meaningful cues may reduce ambiguity. We have also assumed that cue summation–and thereby the reduction of ambiguity–may be obstructed by conflicting image effects on separate attitude scales: and in subsequent work we must investigate the implications of attitude conflict for the cue summation model generally. In condition F of Experiment 19, for example, responses on the strength scale evidently cued by the notes long shots, were found to conflict with those on the straightforwardness scale cued by the location shots. Not as yet discussed, however, is the possibility that individually meaningful cues may have conflicting effects not just upon separate attitude scales but in tandem upon the same scale.

It is a possibility fully recognised by Kaplan (1972). In an important critique of the assumptions underlying the semantic differential approach to attitude measurement, Kaplan indicates SD's inability to distinguish between psychological indifference to a stimulus and ambivalence, the tendency towards simultaneous positive and negative attitudes definable, quite simply, as 'mixed feelings' (Brown, 1965). No two attitudes to a particular stimulus are mutually exclusive, suggests Kaplan—not even those directly opposed to one another such as GOOD and BAD, STRONG and WEAK. For each may arise from distinctive elements within the stimulus; and, when attitudes to any complex event are measured, subjects should therefore be allowed to judge its potentially conflicting elements independently via separate monopolar scales. Future researchers must certainly consider this recommendation carefully in choosing the measurement device most suited to their purpose.

If the experimental procedures had allowed subjects to rate opposing stimulus elements independently in the present research, it is possible that much useful information regarding response ambivalences on individual attitude scales would have been obtained. As it is, however, any instance of bipolar conflict in the viewers' judgements has presumably been expressed in the form of an intermediate rating, and the evidence for it has thereby been concealed. Fortunately, sufficient detail of image effects has so far emerged in general for speculations and predictions regarding their ambiguities as well as consistencies to be made quite adequately. Thus, in Experiment 6, each of the meaningful image cues has appeared to affect quite separate rating scales; and consequently, in Experiment 19, various conflicts between these cues have been evident on an interscalar basis. In Experiment 7 on the other hand, featuring the schoolchildren as subjects, a wider range of meaningful cues has been found to influence overlapping sets of scales; and based on these effects we have inferred a far greater responsiveness to, and tolerance of, visual ambiguity on the children's part than so far highlighted in the present chapter. We have suggested indeed that the children may well have been ambivalent in their responses to separate aspects of particular image cues, as per Kaplan's thesis (1972). At this point, however, the measurement tool we have used throughout this research ceases to be reliable for all purposes; and if we wish to isolate the rules of cue summation in situations where attitude conflicts are substantial we must henceforward adopt an experimental technique of the type that Kaplan has encouraged.

Meanwhile, the existing procedure for the exploration of summative effects as used in the last experiment has been applied to the results of Experiment 7, and the same curvilinear trend has been noted as above albeit on the basis of different cued responses and a different order of preference for the various conditions in behavioural terms. The pronounced between-condition differences in the children's responses on virtually every one of the measurement scales used leaves no doubt as to the systematic nature of attitudinal and behavioural image effects upon these subjects, or to the predictive role of cue summation theory in their examination. The results of Experiment 8, however, yielded by the clerical staff are by no means as systematic. The meaningfulness of particular cues for these subjects was evidently insufficient as a basis for statistically significant summative effects. The above conclusions are based on the behavioural results reported in Experiments 21 and 22.

Experiment 21: Behavioural effects–BVT (children)

As a test of the relationship between attitudinal and 'donation' responses apparent thus far, the earlier study of children's reactions to the same TV material (Experiment 7) is now revisited, as Experiment 6 was revisited in Experiment 18. The approach used in Experiment 18 is reapplied:

1. The image cues identified with significant differences between subjects' reactions to particular presentation conditions are established.
2. The conditions containing the most and least auspicious combinations of cues respectively are determined.
3. The correspondence between these two conditions and those evidently the most and least effective on the donation measure is inspected.

Procedure

As in Experiment 18, the donation measure was administered following the semantic differential measurement of subjects' reactions to the six 'British Vagrancy Trust' presentations. The subjects on this occasion were 74 schoolchildren aged between fourteen and seventeen years.

Results

The amounts of money notionally donated to the six presentation conditions were subjected to a two way analysis of variance, indicating a significant difference between two or more levels of visual complexity ($F' = 6.18$; $df = 2, 68$; $P < 0.01$). The presentation receiving on average the highest donation, £40.80, was F (Profile + Reaction + Location). Multiple comparisons of the conditions revealed that F was significantly more efficient in this respect ($P < 0.01$) than C (Direct + Notes), the least efficient money raiser receiving an average of £7.71; F was also more efficient ($P < 0.05$) than condition D, recipient of an average £13.50. C in turn was significantly less efficient, at the 1 per cent level, than conditions A (£26.21), B (£25.91) and E (£27.54).

From the report of Experiment 7 it will be recalled that all four of the image variables manipulated in these conditions were associated with significant effects upon the children's attitude ratings. The major variables proved to be the added location details and notes long shots, as with the students. In addition the camera angle and interviewer reaction variables yielded significant though ambiguous effects. By contrast with the students' reactions, however, the addition of the notes long shots in condition C was identified with less favourable attitude ratings than those given in the 'no notes' condition A (Table 2.4); this was the case on 9/10 attitude scales significantly differentiated by this manipulation. By the children, therefore, 'no notes' is evidently regarded as a more favourable state than 'notes'; and the condition combining both of the unambiguous positive characteristics (Notes – , and Location +) is F. The condition which is the converse of F in these terms, containing notes detail though no location shots, is C. Even without the independent evidence of the donation measure conditions F and C may thus be predicted, by the same criteria as applied in Experiment 18, to exert the most and least favourable general impacts respectively. The evidence of the donation measure corroborates this prediction.

Discussion

Ratings on the donation measure by schoolchildren were at their most positive when the significant effects of image cues upon their attitudes summated.

The reactions to the appeal furnished by schoolchildren are thus very similar to those of the students in Experiment 18, firstly in that the image cues producing unambiguous attitude effects are common to both subject groups: also in that the reactions to these cues successfully predict the extreme effects on the donation measure. As noted previously, however, other characteristics of the image effects on children and students differ. In the students' data the influential image cues were found to operate upon individual attitude scales with a high degree of selectivity. In the present context, however, a wide range of significant effects has been observed on far more of the attitude scales overall. The added location shots particularly, in conditions E and F, seem to have exerted a quite unselective impact on many scales at once, indicating a simple 'positive/negative' evaluation; and the predictive relationships between the attitudinal and behavioural effects in this experiment has been evident without recourse to the highly discriminating technique of factor analysis as on the previous occasion.

We may predict, however, that the children's reactions to the very detailed visual conditions E and F will have a far simpler evaluative structure than those given in other conditions. Factor analyses of these data reported in Chapter 5 reveal this to be the case. The full question of image effects upon judgement structure is taken up in that context in the light of factor analyses of the total attitudinal data reported to this point.

Experiment 22: Behavioural effects–BVT (clerical staff)

A further test of the predictive relationship between attitudinal and behavioural effects is now made using the data from Experiment 3. Also featuring attitude ratings of the 'British Vagrancy Trust' material, this experiment provided a final check within Chapter 2 on the susceptibility to image effects of different types of subjects.

Procedure

The subjects on this occasion were 66 clerical workers: see Experiment 8. Following the measurement of their attitudes in each 'BVT' condition, the donation measure was administered as in Experiments 18 and 21.

Results

A two way analysis of variance was performed upon the amounts of notional money donated in the six conditions, yielding no significant between condition differences. The average donations ranged from £9.60 in condition E (Direct + Notes + Location) to £16.00 in condition D. (Additionally: A = £10.45; B = £15.73; C = £15.00; F = £11.00.) Notwithstanding the lack of significant differentiation, it is worth noting that the condition receiving the highest donation (D) was also that combining the two image cues (Profile + Reaction) associated with significant positive effects upon the clerical staff's attitude ratings (see Table 2.7).

Discussion

Ratings on the donation measure by clerical staff were at their most positive when the significant effects of image cues upon their attitudes summated, though the between-condition donation differences were not significant statistically.

As already seen in Experiment 8 the significant impact of the camera angle and interviewer reaction variables on these subjects involved two of the attitude scales only; RELAXED/TENSE and EXPERT/INEXPERT. While these effects may certainly relate to marginal differences between the conditions on the behavioural measure, accounting for the slight if non-significant superiority of condition D as a notional money raiser compared with the other conditions, the results are clearly insufficient as evidence for a predictive relationship between attitudes and behaviour generally. It is apparent that not all subject types are susceptible to behavioural effects as measured on this basis. Just as the students appear less susceptible to presentation effects than the children, possibly being more critically aware of their irrelevance to the judgement tasks at hand, it appears that the clerical workers were less susceptible to visual persuasion than both types of subject. Intuitively one might have expected the highly intelligent undergraduate students to be the most resilient subjects in this respect; though of the three types of subject investigated here it is the clerical workers who may perhaps be considered the most avid and experienced of TV viewers. University students are notoriously less avid than other sections of the viewing population, and they may thus be relatively more susceptible to visual effects than one might suppose. The questions of visual literacy and its various levels suggested by these comparisons merit much further examination.

Experiment 23: Presentation and immediate recall

The only attempt so far made in the course of this research to establish the effects of TV presentation on a further psychological measure was unsuccessful. In emphasising the differential effects of presentation technique upon attitude measures we have consciously given little attention to the possibility that subjects' recall for televised material may be affected also, or to the development of methods for testing it. The problems of memory testing are no less formidable than those of attitude measurement; and our decision to devote the substantial part of this research to the latter was based on the faith that attitude measures have greater predictive potential as indicated in Chapter 1. Learning gain from TV has in any case been the subject of extensive research previously. Directly relevant to the present research are studies of the general beneficial effects of visual presentation upon learning by Booth (1970-71), Edwardson et al. (1976), Katz et al. (1977) and by Salomon and Cohen (1977). News items featuring moving film have been shown to facilitate greater recall of information than comparable items illustrated by photographic stills only (Gunter, 1979): cf. the similar result of our own Experiment 12. Findahl and Hoijer (1977), however, have indicated that the general evidence relating to comparisons of film and still presentation techniques is inconclusive. Clearly a prime reason for the latter is the lack of general correspondence between studies as regards methodology; and, as in Experiment 12, we conclude that a standardisation of procedures for the examination of image effects is now needed.

With respect to the learning and recall effects of TV presentation, the work of the Children's Television Workshop (Palmer, 1969) and that of Findahl et al. in the Audience Research Department of Sveriges Broadcasting Corporation (1972, 1976, 1977) has certainly been the most concerted. The Swedish work in particular parallels our own: it has considered the impact of visual as well as verbal techniques in news broadcasting, and has carefully discriminated between viewers' recall for the events, locations and persons covered by news items and for the items' consequences. It also has the advantage of its association with a large scale news broadcasting enterprise, and in these general respects it provides a useful model for future investigators to emulate. Unfortunately, the statistical treatments reported by the Swedish team involve few tests of significance, and some of their results are therefore difficult to evaluate. However, their ultimate conclusions are beyond doubt: that verbal content, repetitions of it, and the visual information accompanying it must be carefully designed to be mutually complementary (Findahl et al., 1976, pp.33-41); moreover, (1977, pp.41-2):

> that the 'professional' journalistic way of seeing things is not always the most appropriate, and that a discussion of the various means of achieving clarity must be carried on continuously.

Regarding effects of repetition and review strategy upon attitudinal as well as recall measures, see also Coldevin (1975a, b).

The present work attempts to serve as one cornerstone in this discussion, providing methods for use in the attitude area of image effects specifically. As a comprehensive review of previous research by Coldevin (1976) has indicated, the effects of presentation on attitudes have traditionally been far more amenable to measurement than recall effects; and, as argued in the present chapter, they relate to higher order psychological effects on a predictive basis. The majority of previous studies has either contrasted a number of different presentations, requiring viewers to recall as many of the items involved as possible, or has given pre- and post-tests on a range of highly specific presented facts. The types of text used in the present experiments rule out the first of these strategies and are generally too brief for the second. The 'British Vagrancy Trust' material alone has been thought possibly exceptional in this respect, lasting over five minutes and permitting a detailed post-test of subjects' recall for the verbal material contained within it.

Procedure

Following testing of the students' reactions to the 'BVT' material in Experiment 6, therefore, a questionnaire was administered containing 30 phrases taken from the appeal's text. Up to four words were omitted from each phrase and the 60 subjects were instructed to 'fill in the gaps from memory as literally as you possibly can'. The number of missing words was specified in each case. One point was subsequently awarded for each word correctly recalled even though its position in the sequence of words may have been recalled wrongly. An equal number of items was derived from those sections of the text which, in the visually elaborate conditions E and F, had/had not been accompanied by added location detail. These two classes of item (visual and non-visual) were also balanced for difficulty in terms of the number of words to be recalled.

Results

A two way analysis of variance indicated that the amounts recalled on this basis in the six presentation conditions were not significantly different. Further t-tests intended to establish whether, in conditions E and F, verbal retention was aided or indeed impeded by simultaneously presented location detail, elicited no significant differences either.

Discussion

Students' immediate recollections for TV material in various presentation conditions were not significantly different.

The questionnaire technique used in this instance may have been too stringent or too simple as a recall index. Moreover, it should be noted that the 'BVT' text itself was not essentially fact laden in the manner of material ideally suited to a memory test. In reporting a similarly unsuccessful attempt to relate presentation variables to learning gain, Coldevin (1977) has suggested that they may be less critical as determinants of factual recall than in the influencing of opinion change. If this should henceforward prove to be the case, the general effects of image variation will probably be more apparent on delayed rather than immediate recall bases; for, as Aronson (1973) has demonstrated, the influences of a message upon its audience are often not apparent until some considerable time after its reception. In the present research the testing of delayed recall effects has been neither convenient logistically nor practicable without our sacrificing the provision for subjects to give their responses anonymously.

No further attempts to gauge the impact of TV presentation upon recall measures have therefore been made during the present research, though in view of the particularly high susceptibility to image effects shown by the children in Experiments 7 and 21, the question should certainly be investigated with reference to different types of viewer—and indeed to the possible effects of cue summation. The attention given to retention and recall measurement in the ongoing studies of TV presentation impact by Sullivan et al. (1977, 1978, 1979) represents a useful step in this direction. Their research, indicating that student attitudes towards TV material—pro or con—tend to intensify over time, has already supported the notion of TV's 'sleeper' effects mentioned above. They have confirmed, moreover, via significant correlations between the different measures used, that attitudinal data can serve a valuable predictive function in the isolation of TV's instructional effects generally. At this stage we therefore call a halt to our own investigations of TV's effects upon quasi-behavioural measures and continue by assessing the overall evidence gathered so far concerning image effects upon attitude formation and conflict.

5 Interpreting image effects

This chapter reappraises the attitude effects reported to this point and considers the major image variables evidently causing them. If the TV image is indeed host to a range of variables systematically influencing audience judgements as the earlier experiments suggest, the fact should be reflected in the systematic variation of the dimensions on which viewers' attitudes are based in different conditions. Having found factor analysis to be of use as a means of identifying the structure of viewers' judgements in Experiments 19 and 20, we apply the technique to our data as a whole. In the process we search for clues to the nature of attitude effects taking place.

Throughout the experiments it has appeared that many of these effects involve subjects in a degree of attitude conflict. For instance, in Experiment 13 the performer was noted as appearing more RELAXED and yet also more RASH when using the autocue device a great deal, as opposed to seldom; and in Experiment 10 a performer seen to be giving a 'positive' political message was rated as certainly more POSITIVE than when associated with a 'negative' message but also less RELIABLE. Similarly, in the TV appeal situation of Experiment 17, the presentation style regarded as the most PLEASANT was also significantly the least effective in terms of the behavioural 'donation' measure. In highly explicit perceptual situations it is evident from the vast literature devoted to semantic differential applications (Snider and Osgood, 1969) that bipolar rating scales are likely to be used primarily on a very simple 'evaluative' basis: the majority of responses reflects whether the subject is either for or against the object of attention and does not observe more minute shades of meaning between the scales. Responses to the TV situations mentioned above, however, are clearly not restricted to a simple 'positive-negative' dimension such as this. Accordingly we now apply factor analysis in an attempt to determine whether particular viewing conditions generate different levels of meaning which, if not independent, may actually conflict with one another to produce the types of attitude effect we have noted.

The factor analytic method most appropriate for the purpose of highlighting the individual levels of meaning within a group's attitude data involves—unlike in Chapter 4—a rotational procedure. Following factor rotation the clusters of scales that a group's members consistently use in a similar manner—presumably because they reflect similar qualities such as 'pleasantness', 'friendliness', 'niceness', and so on—are emphasised. A group's responses on a set of 20 attitude scales may be reduced to perhaps five or six separate dimensions in this manner, each one reflecting independent judgement criteria being applied. As indicated previously, it is common for judgements of professional communicators to yield prominent 'potency' factors (cf. Tannenbaum, 1955; Mandell and Shaw, 1973). More specific qualities commonly defining communicative prowess are, as Kelley (1967, p.204) affirms, perceived trustworthiness and expertise. We check this observation in the experiment that follows.

In different communication situations, however, such factors may have a wide range of connotations. 'Expertness', as defined by Kelley (p.204), can variously suggest a communicator's command over his information or his ability to mediate it to others—two separate capacities clearly capable of quite independent assessment by an audience. The expertise of individual communicators may be appropriately judged in terms of either or indeed both capacities. A newscaster, for example, not necessarily assumed to have any personal knowledge of the information he conveys, may be judged as thoroughly expert in terms of a confident style of delivery alone. A degree of formality in his delivery may enhance his perceived expertise while detracting from that of a TV interviewer. The skills of communication, in short, are defined (a) in terms of the functions the communicator fulfills, and (b) according to his ability to cope with the environment in which he must perform. In seeking to establish the criteria underlying performer assessments, therefore, we must recognise that they have probably been influenced as much by variables such as these as by the others acknowledged in the experimental design. Criteria found to underly subjects' judgements in any one experimental condition may certainly not be assumed to predict those operating within others, unless independent evidence of their general relevance across conditions has become available.

Experiment 24: Analysis of judgement criteria

The first step in an attempt to isolate and interpret the system (if any) underlying our earlier data must thus be to seek for evidence of general judgement criteria of this type. We wish to know whether particular rating scales have been used throughout the experimental series with a constancy demonstrating their importance for the subjects at large regardless of conditional variations. The present study was conducted accordingly.

Procedure

The data collected in Experiments 1 to 17 were amassed. In total these experiments have investigated audience reactions in 66 experimental conditions. Contrasting types of TV performance have been studied in a wide range of TV formats. The ratings elicited in each individual condition were now factor analysed via the principal factoring technique with iteration previously applied in Experiment 19; the number of factors extracted in each condition was once again determined by the 'eigenvalue one' criterion, and the initial solution was rotated by the Varimax method (Kaiser, 1956) so that the distinct groups of interrelated attitude scales might be emphasised.

Each scale was now inspected in terms of its tendency, in the different rotated solutions, to cluster with each of the other scales. The minimum loading regarded as denoting a significant relationship between the scales and each rotated factor was, as before, set at plus or minus 0.55. It is immediately apparent from Table 5.1 that certain scales (particularly those measuring assessments of RELIABLE/ UNRELIABLE and SINCERE/INSINCERE) were frequently found to share high factor loadings with other scales. It seems that particular between-scale relationships persist despite the considerable variety of experimental conditions in which they are observed. It also emerges that certain scales are never related on this basis at all: assessments of performers' expertise and humanity, for example, invariably load highly on separate rotated factors and are clearly formed quite independently throughout the range of viewing situations we have examined. Indeed several tendencies are noted for high positive judgements on one scale (e.g. STRONG/ WEAK) to be related to high negative ratings on another (e.g. RUTHLESS/ HUMANE)—significant inverse relationships suggesting conflict between separate judgement criteria. The table gives the frequencies of association for each pairing of the 19 judgement scales most commonly related to other scales in the rotated factor solutions: for brevity their positive pole only is specified. The remaining scales (not quoted) have distinctively lower association totals, owing either to their less frequent usage during the experimental series or to their less systematic treatment by the subjects. The threshold between high and low association totals was determined by the 'scree test' (Cattell, 1966a, Rummel, 1970).

It must be stressed that the individual frequencies in Table 5.1 do not necessarily express tendencies towards a significant selective relationship between scales. The high number of associations between the reliability and sincerity scales, for example, may be solely due to the high tendencies of each to associate with other scales generally. Such scales may, for instance, be used as indices of a general evaluative judgement either for or against TV performers in the classical Osgoodian sense. For clear evidence of the system actually underlying viewers' attitudes in the 66 conditions, the association matrix must now be factor analysed in its own right.

In presenting the association data for factor analysis the frequencies of inverse scalar relationships, where noted, were subtracted from those of the corresponding positive relationships. Being already collated in the symmetrical matrix form generally not obtained until after the correlation stage of a factor analysis, the data were next 'bounded'—i.e. transformed into simulated correlation coefficients via the division of all values by 11, the largest frequency in the matrix (Rummel, 1970, p.293). The cells representing each scale's 'relationship' with itself (the communality cells falling on the main diagonal of the matrix) were set at unity; and the whole was principal factored using a non-iterative technique in view of the relatively large number of scales involved (cf. Comrey, 1973, p.74).

Results

A two factor solution was first of all obtained, and subjected to Varimax rotation. (As discussed earlier, rotation of factors becomes desirable when the primary clusters of interrelated attitude scales are of interest as opposed to the patterns of interrelationship formed by the data as a whole.) Inspection of the rotated solution in Table 5.2 and Figure 5.1 indicates a major division of the scales into:

1. Those estimating a performer's Professional qualities (his perceived strength, profundity, expertise, interest value and straightforwardness in the communication of information).

2. Those estimating the more Personal characteristics of fairness, humanity, honesty, reliability, pleasantness and sincerity.

(NB The unrotated solution is as in the first two columns of Table 5.3). The two factor system clearly reflects that discussed by Kelley (1967) in terms of expertise and trustworthiness (see previous section).

Table 5.1

Association frequencies of rotated loadings – Experiment 24

Scale		1	2	3	4	5	6	7	8	9	10	11	12	13	14	15	16	17	18	19
RELIABLE	(1)	(0)	8	9	8	6	8	3	5	0	3	5	2	2	0	2	1	1	1/2	1
SINCERE	(2)	8	(0)	8/1	7	3	6	5	5	3	4	4	2	3	0	0	1	2	0	1
HONEST	(3)	9	8/1	(0)	6	2	2	2	3	2	5/2	4	4	1	0	2	1	4	1	1
FAIR	(4)	8	7	6	(0)	3	2/1	2/1	3	5/1	4	6	6	3	1	2	2	1	1	0/1
DIRECT	(5)	6	3	2	3	(0)	5	2	4	2	4	0	0	1	1	1	0	3	0	0
EXPERT	(6)	8	6	2	2/1	5	(0)	6	9	4	7/1	0	2/1	0	2	2	0	3	1/1	2
PROFOUND	(7)	3	5	2	2/1	2	6	(0)	11	9	5	2	4	2	1	1	2	1	1	3
STRONG	(8)	5	5	3	3	4	9	11	(0)	5	5	1/3	0	1	0	4	4	1	1	1
INTERESTING	(9)	0	3	2	5/1	2	4	9	5	(0)	5	3	4/1	3	1	0	5	0	1	3
S'FORWARD	(10)	3	4	5/2	4	4	7/1	5	5	5	(0)	0	2	1	0	0	1	0/1	2	0
HUMANE	(11)	5	4	4	6	0	0	2	1/3	3	0	(0)	7/1	3	0	2	1	0	2	1
PLEASANT	(12)	2	2	4	6	0	2/1	4	0	4/1	2	7/1	(0)	3	2	1	0/1	0	2	0
FRIENDLY	(13)	2	3	1	3	1	0	2	1	3	1	3	3	(0)	0	1/1	3	0	3	1
NOT NERVOUS	(14)	0	0	0	1	1	2	1	0	1	0	0	2	0	(0)	2	0	1	0	0
RELAXED	(15)	2	0	2	2	1	2	1	4	0	0	2	1	1/1	2	(0)	1	0	0	0
PERSUASIVE	(16)	1	1	1	2	0	0	2	4	5	1	1	0/1	3	0	1	(0)	0	0	0
BELIEVING	(17)	1	2	4	1	3	3	1	1	0	0/1	0	0	0	1	0	0	(0)	0	0
POPULAR	(18)	1/2	0	1	1	0	1/1	1	1	1	2	2	2	3	0	0	0	0	(0)	2
IMPORTANT	(19)	1	1	1	0/1	0	2	3	1	3	0	1	0	1	0	0	0	0	2	(0)
Association totals	+	65	62	57	62	37	61	62	63	55	48	41	41	31	11	21	22	17	18	16
	–	2	1	3	4	0	4	1	3	2	4	4	4	1	0	1	1	1	3	1

(NB Where two figures are given—e.g. 1/3—the second represents the total number of inverse relationships observed between scales)

Table 5.2

2-factor (rotated) solution of Table 5.1 – Experiment 24

Scale	*Factors* I	II
RELIABLE	.421	(.735)
SINCERE	.474	(.639)
HONEST	.215	(.750)
FAIR	.161	(.863)
DIRECT	.466	.202
EXPERT	(.842)	.121
PROFOUND	(.869)	.104
STRONG	(.967)	.027
INTERESTING	(.640)	.180
S'FORWARD	(.615)	.185
HUMANE	-.115	(.781)
PLEASANT	.010	(.640)
FRIENDLY	.095	.388
NOT NERVOUS	.098	.060
RELAXED	.176	.167
PERSUASIVE	.296	.064
BELIEVING	.151	.150
POPULAR	.043	.167
IMPORTANT	.240	-.002
Total variance*	21.6%	19.2%
Common variance*	53.0%	47.0%
Interpretation	Professional	Personal

(NB Loadings greater than ± 0.50 in parentheses)

*Variances estimated from data taken to five decimal places

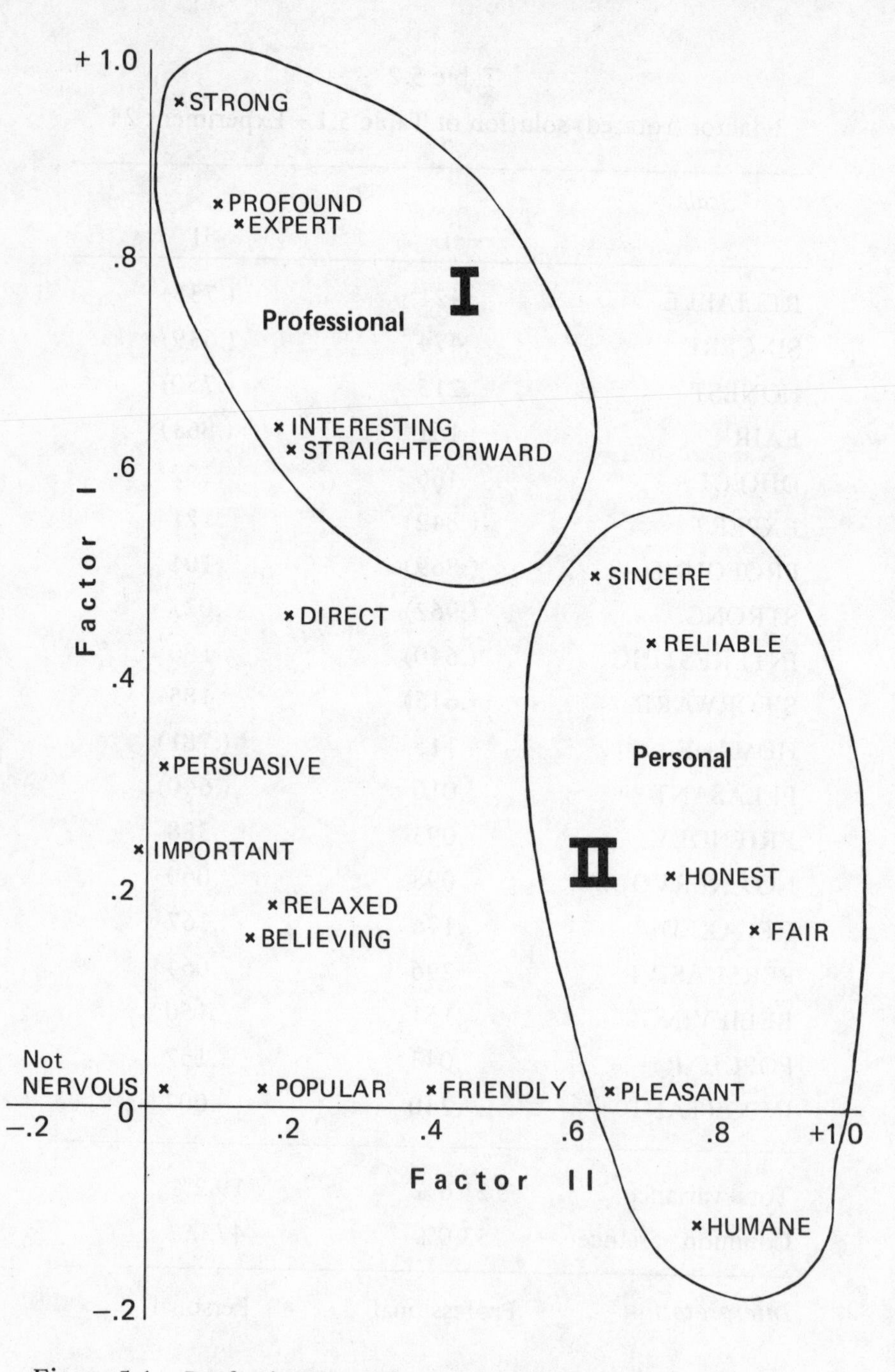

Figure 5.1 Professional and Personal attitude factors – Experiment 24 (NB See Table 5.2)

However, when four principal factors were extracted from the data by the same method an even clearer system was observed (Table 5.3). The four (rotated) factors are clearly identifiable thus-

1 A pronounced relationship is noted between the reliability, sincerity, honesty, directness and fairness scales. The high positive loading of these scales on the same rotated factor in this solution suggests a tendency for viewers, in a variety of conditions, to assess the performers' personal Integrity.

2 An independent tendency is evident towards the relationship of profundity, strength, interest value and straightforwardness assessments: i.e. those expressing a performer's professional Mastery of his information and its organisation.

(NB One scale—that concerning perceived expertise—is related to both of the above factors: it has, in other words, a dual meaning or 'variable complexity' of 2, cf. Rummel, 1970, p.325.)

3 A third factor is identified with the humanity, pleasantness and friendliness scales, each reflecting a performer's ability to project himself to his audience at a personal level. A similar 'viewer engaging' factor has been extracted from ratings of a TV performer by McMenamin (1974) and labelled Empathy.

(NB The 'fairness' scale has a dual relationship to factors 1 and 3.)

4 The fourth factor indicates a relationship between perceived nervousness and tension: it thus represents a performance quality also identified by McMenamin (1974) as Poise.

While the Mastery and Poise factors reflect levels of skill necessary to the communicator in his professional capacity (as e.g. a capable newsreader or lecturer), Integrity and Empathy concern his ability, judged quite independently by the viewers to inspire personal trust and liking. It may also be argued that the former factor in each pair relates to the innermost, fundamental levels of a communicator's credibility, while the latter represents the evidence of his abilities to utter, or 'outer' himself publicly. This interpretation of the four factors is summarised in Figure 5.2. Between them the factors account for nearly 60 per cent of the total variance in the matrix.

Discussion

In the earlier experiments as a whole, viewers tended to judge TV communicators on independent Professional and Personal bases and on different levels of each.

The distinction between professional and personal criteria for communicator assessment is amply supported by previous evidence in the literature. In an experiment which required students to judge various public speakers on semantic differential scales as here, Andersen (1961) found factor analysis to extract two dimensions from his data—one relating judgements of honesty, morality, fairness and goodness (clearly comparable with our Integrity factor), and the other reflecting perceived interest value, strength, speed and activity (i.e. the Mastery factor). Berlo (1961) confirmed these results. From ratings of speakers both imagined, as in the earlier two studies, and actually presented on tape as in our own, McCroskey (1966) extracted factors interpreted as reflecting 'authoritativeness' and 'character'—again clear professional and personal

Table 5.3

4–factor solutions of Table 5.1 – Experiment 24

Scale	*Unrotated factors* I	II	III	IV	*Rotated factors* I	II	III	IV
RELIABLE	(.798)	.285	-.483	-.136	(.960)	.059	.212	.008
SINCERE	(.776)	.178	-.200	-.236	(.757)	.249	.275	-.136
HONEST	(.651)	.430	-.261	-.069	(.723)	-.016	.398	.019
FAIR	(.683)	(.551)	.000	-.082	(.598)	.048	(.645)	-.038
DIRECT	.485	-.149	-.405	.000	(.586)	.192	-.165	.115
EXPERT	(.718)	-.456	-.321	.160	(.618)	(.570)	-.248	.288
PROFOUND	(.728)	-.486	.355	.119	.184	(.918)	.110	.138
STRONG	(.752)	(-.608)	-.035	-.007	.457	(.825)	-.202	.085
INTERESTING	(.603)	-.280	(.586)	-.042	.006	(.818)	.333	-.076
S'FORWARD	(.588)	-.260	.060	-.115	.355	(.546)	.041	-.063
HUMANE	.420	(.668)	.282	.088	.217	-.048	(.811)	.055
PLEASANT	.424	.479	.368	.385	.080	.119	(.747)	.338
FRIENDLY	.324	.233	.440	-.212	.020	.250	(.517)	-.260
NOT NERVOUS	.114	-.018	.008	(.720)	-.049	.088	.095	(.716)
RELAXED	.242	.012	-.095	.478	.157	.094	.084	(.506)
PERSUASIVE	.267	-.144	.349	-.385	.002	.416	.151	-.408
BELIEVING	.212	.016	-.356	.151	.363	-.053	-.099	.226
POPULAR	.141	.098	.407	.102	-.166	.215	.361	.042
IMPORTANT	.181	-.158	.254	.088	-.079	.332	.098	.065
Total variance	28.4%	12.5%	10.4%	6.7%	19.6%	17.6%	13.8%	7.0%
Common variance	49.0%	21.5%	17.9%	11.6%	33.8%	30.3%	23.8%	12.1%
Interpretation			z		Integrity	Mastery	Empathy	Poise

(NB Loadings greater than ± 0.50 in parentheses)

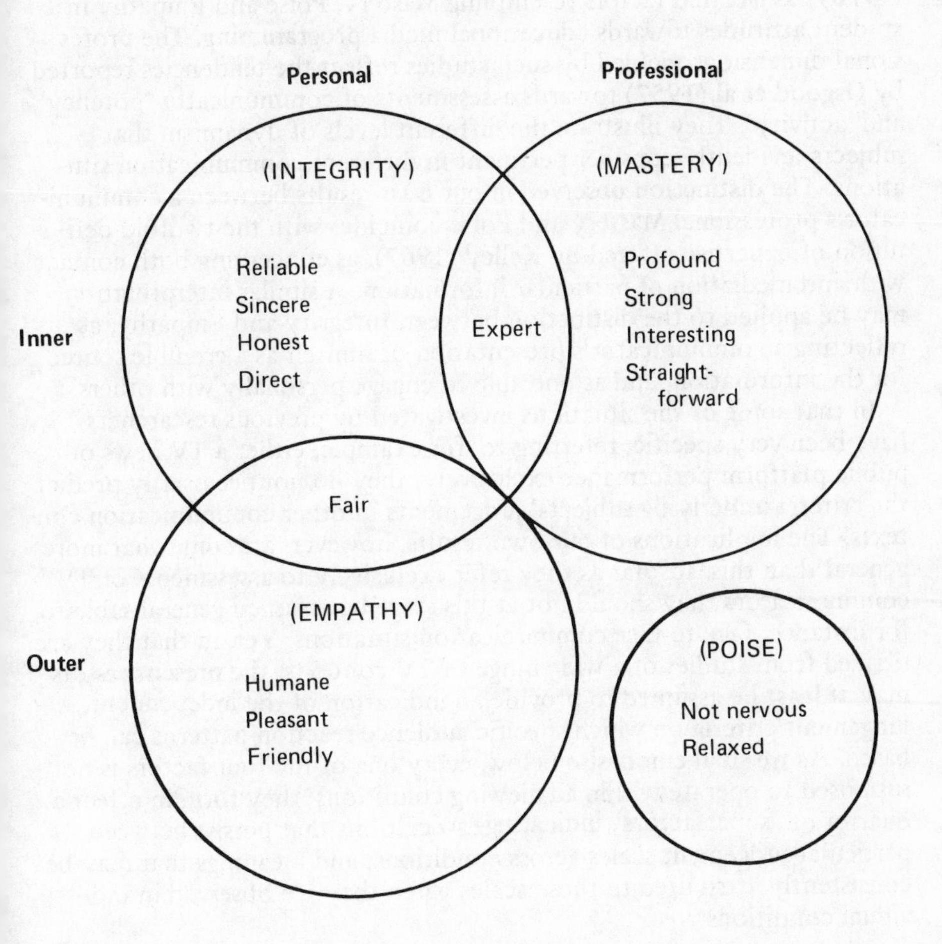

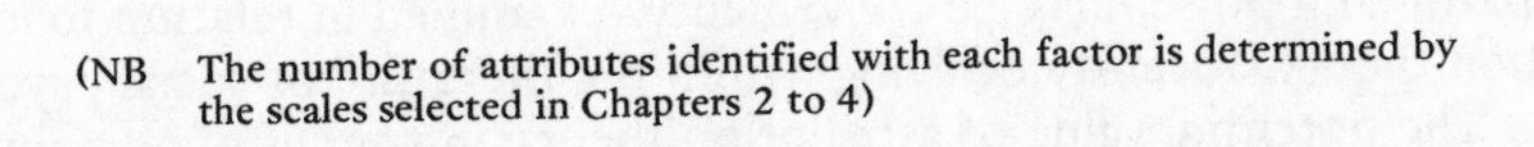
(NB The number of attributes identified with each factor is determined by the scales selected in Chapters 2 to 4)

Figure 5.2 Orthogonal judgement criteria – Experiments 1 to 17

dimensions. Highly comparable results are also reported by Williams (1963), Markham (1968) and McMenamin (1974). Additionally, Morgan (1978) has isolated factors resembling Mastery, Poise and Empathy in student attitudes towards educational media programming. The professional dimensions yielded by such studies reflect the tendencies reported by Osgood et al.(1957) towards assessments of communicator 'potency' and 'activity'. They illustrate the different levels of dynamism that subjects evidently consider pertinent in different communication situations. The distinction observed in our own results between a communicator's professional Mastery and Poise coincides with the twofold definition of expertise offered by Kelley (1967), as concerning both contact with and mediation of particular information. A similar interpretation may be applied to the distinction between Integrity and Empathy, as reflecting a communicator's presentation of himself as a credible source for the information, and as one able to engage personally with others.

In that some of the situations investigated by previous researchers have been very specific, referring to, for example, either a TV news or public platform performance exclusively, they do not necessarily predict the criteria underlying subjects' judgements in other communication contexts. The implications of our own results, however, are somewhat more general than this. Insofar as they refer exclusively to assessments of TV communicators they should not at this stage be assumed generalisable to, for instance, face to face communication situations. Yet, in that they are derived from studies of a wide range of TV contexts, the present results may at least be assumed to provide an indication of the independent judgement criteria on which specific audience reaction patterns can be based. As we shall emphasise below, every one of the four factors is not supposed to operate within all viewing conditions; they form in effect a quartet of 'superfactors' indicating associations that persist between particular judgement scales across conditions, and meanings that may be consistently attributed to those scales when they are observed in individual conditions.

Attribution theory and attitude patterns

The indication that, despite variations in the viewing condition, a TV audience tends to judge performers in terms of their mastery, integrity, empathy and poise, may be conveniently examined in relation to attribution theory, recently evolving within the literature of social psychology. The potential value of attribution theory in studies of communication and persuasion has been indicated by Kelley (1967). The theory regards psychological and behavioural processes as essentially determined by man's desire to interpret his environment, attributing causes to the events and qualities he perceives there so that he may come to predict and control them (cf. Kelley, p.193). It is clearly consistent, therefore, with the 'personal construct' approach to psychological theory recommended by the other Kelly (1955), and applied in the TV research context by Baggaley and Duck (1976). The process of attribution itself is regarded as an attempt to render the factors underlying one's environment into a meaningful, structured form; and yet this only proves possible in terms of the 'variable manifold' of events by which the environment is mediated. The origins of the latter notion, central to attribution theory, may be traced in the work of Heider (1958).

The effects of TV imagery can be readily interpreted in these terms. The four judgement factors identified in the previous section are held to represent the comprehensive set of criteria available to viewers in forming attributions about a TV communicator: not all criteria are necessarily appropriate in every viewing context. Similarly the 'variable manifold of mediating events' à la Heider embraces all sources of information that a viewer may, rightly or wrongly, consult in seeking to structure the impressions he forms. Not only can the manifest qualities of the performer himself be used in this connection, but also, as Kelley (p.194) indicates, the reactions of other persons to him and connotations arising from the modality or medium via which he is presented. The effects of camera angle, visual detail and editing procedure clearly arise on the latter basis; they are examples of the biases, errors and illusions in attribution due to causal factors of which the subject may be quite unaware (Kelley, pp.219-35). In practice, however, it is important to recognise that the sum total of a subject's responses to a perceptual situation may be the result of information simultaneously derived from a wide range of sources, and weighed one item against the other as efficiently as possible. Subjects have the often awkward task of 'disentangling which effects are to be attributed to which of several factors present' (Kelley, p.194), an endeavour identical to that of the systematising scientist (see Chapter 7).

As long as the factors referred to by a given subject yield complementary information any attributions simultaneously based upon them may be expected to coalesce in an harmonious manner evident from the consistency of the subject's responses. When psychological effects upon a group of subjects are measured their response tendencies may be estimated in terms of the patterns indicated by factor analysis. In the present context attitude tendencies which regularly agree with one another will form, for the group of subjects evincing them, an harmonious or 'consonant' attitude pattern denoting an orientation towards, for example, a performer's personal qualities, or his professionalism, or any one of the four independent criteria for communicator assessment highlighted by Experiment 24. By reference to these four elemental criteria we can break down or 'parse' the wide variety of attitude factors which TV conditions yield, in attempting to gain access to the levels of complexity at which reactions to TV are commonly formed,and to the types of response which may be expected in particular conditions.

As Rummel (1970, Chapter 14) indicates, factor complexity is a concept central not only to the interpretation of existing research results but also in the design of future experimentation. It is the complement of 'variable complexity' illustrated in Experiment 24, and refers, at three levels, to the number of variables with which an individual factor is identified. A factor may be either:

1 General, containing moderate or high loadings for many types of variable, and delineating a broad (highly complex) pattern of data relationships;

2 Group, containing moderate or high loadings for a restricted number of variable types (fairly complex pattern); or

3 Specific, containing high loading(s) of one type only, and delineating a very selective pattern of data relationships (cf. Rummel, pp.325-7).,-

In research contexts such as our own, the degree of a factor's complexity indicates the extent of its predictive value both from one experimental context to others and in the attempt to generalise from the empirical situation to 'real life'. Thus, when audience attitudes are simultaneously based on three or four of the Integrity/Mastery/Empathy/Poise criteria, the effect is revealed factor analytically by a general 'evaluative' pattern of the classical type. Yet in relation to particular attribution processes general factors usually have limited predictive value: they may represent various, though highly correlated, criteria that the analysis has failed to separate, or they may derive from an uncritical usage of the rating scales by subjects. The more specific an attitude factor on the other hand, the more it is likely to reflect the impact of distinctive stimulus properties and to suggest hypotheses for future testing. By attention to the complexity/specificity of the factor patterns obtained in each of the 66 conditions analysed during Experiment 24, for instance, it may prove possible–depending on the experimental design relating to each–to trace particular types of pattern to specific variations in the mediating image. We check this possibility, considering the meanings and complexity of attitude factors extracted from our 66 viewing conditions in turn, in the remaining sections of this chapter.

In this connection it is the unrotated factor matrix produced by each analysis to which we must attend, as in Chapter 4. In order to establish whether particular image variations–and by implication the production procedures responsible for them—have recognisable effects upon attitude patterns in specific situations, we shall need to compare the importance of attitude factors in different presentation conditions, also the varieties of harmony or conflict characterising them. As shown in Experiment 19 the importance of individual factors within a solution may be estimated in terms of the percentage of variance in the data for which they account. Once the factors have been rotated, however, this evidence is usually masked, and the emphasis shifts in any case from the general data patterns with which we are at present concerned to the more specific clusters of interrelationship of the type we were concerned to isolate in Experiment 24. The manner in which we interpret the unrotated attitude factors is as follows.

The criterion chosen to denote a significant association between a factor and the individual attitude variables is plus or minus 0.55, as previously. While the specific factors among our data are identified as those representing judgements based on one of the four (I/M/E/P) criteria alone, the more complex factors are seen as those on which judgements based on more than one criterion correlate. Thus at the intermediate or group factor level we identify the patterns revealing a significant relationship between judgements of Mastery and Poise—a grouping of criteria into a moderately complex pattern though specific at least in its orientation towards perceived professionalism; and at the same level we identify the patterns revealing relationships between Integrity and Empathy—a grouping which indicates a moderately complex orientation towards the performer's personal qualities. Similar group factors may be defined distinguishing between inner and outermost qualities (see Figure 5.2). The general factor level (at which attitude patterns yield a more diffuse set of meanings) embraces a wide range of relationships between the different types of group factor, and broad (i.e. 'evaluative') assessments involving judgement patterns not classifiable more specifically. The variation of attitude patterns from the general to specific in this manner is illustrated in Figure 5.3. (Each of the complex patterns may be either consonant or dissonant: see next section.)

In interpreting our attitude patterns in the above terms, we actually avoid a number of descriptive problems that the more traditional E-P-A terminology leaves unresolved. As Osgood et al. have themselves recognised in proposing that attitude tendencies be examined in basic evaluative, potency and activity terms, these three dimensions are by no means guaranteed mutually exclusive. In different situations individual attitude variables may prove equally related to any two dimensions, or indeed to all three of them. Higher order analyses of the scales defined as 'evaluative' reveal that an evaluative factor can itself subsume variables closely related to those others (i.e. potency and activity) supposed to be independent of it (Osgood et al., 1957, pp.70-1): '. . . it was apparent to us that the

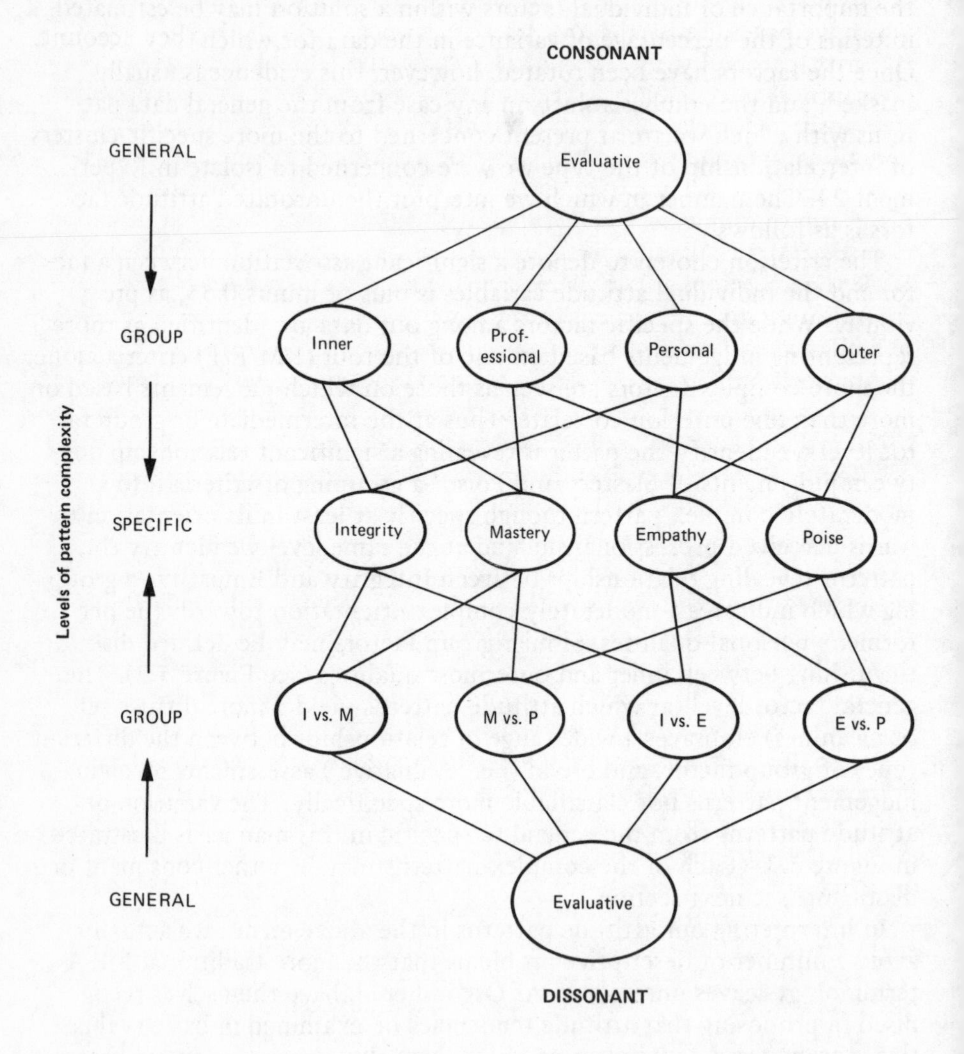

Figure 5.3 Classification of attitude patterns

evaluative dimension . . . was a very general one – a sort of sheath with leaves unfolding toward various other directions of the total space'. In the current research context it is therefore of particular interest that the four 'specific evaluative factors' identified by Osgood and his colleagues at this point closely resemble the four orthogonal judgement criteria which we have discussed above. Firstly, a 'morally evaluative' factor (cf. Integrity); secondly, an 'aesthetically evaluative' factor measuring pleasantness, niceness, etc. (cf. Empathy); thirdly, a factor bearing 'sizeable positive correlations with STRONG' (cf. Mastery) and finally, an 'emotionally evaluative' factor measuring calmness, relaxedness, etc. (cf. Poise). As a vocabulary for the basic, elemental criteria underlying attitude ratings in universal psychological contexts, therefore, it appears that the I/M/E/P distinctions have a solid foundation in previous theory, while the variations in the manner of their usage in specific conditions are indicated via the language of factor complexity.

Evidence for the general and specific tendencies underlying audience attributions may naturally serve to generate insights and hypotheses concerning media effects of numerous types. The hoary question of the social influence of TV violence, for example, would certainly be assisted if specific types of attitude consonance or dissonance were henceforward found to be produced in those who view it. For practical purposes the task would then remain to trace these effects to the particular production elements causing them—an effort requiring the careful identification of individual production variables beforehand. Of course, in any such inquiry particular psychological tendencies can accidently be highlighted or overlooked if any one of the criteria on which they are based is not represented at the data collection stage. Appendix 3 (obtainable from the author) reveals that during the present series of experiments this has occasionally happened with reference to the judgement criterion of Poise, relatively underrepresented in the research as a whole and only reliably predicted by ratings of perceived relaxedness/tension and nervousness. Fortunately, the psychological importance of the Poise criterion and its sensitivity to the effects of image variation are amply evident in the data as a whole, while the manner of its frequent interaction with other judgement criteria in individual viewing conditions becomes apparent in the analyses reported below.

Mediation and attitude complexity

With the exception of Experiment 9, in which more than one image variable per condition was manipulated, each of Experiments 1 to 17 yields several contexts in which the relationships between presentation procedures and attitude factors may now be studied in detail. In this section we illustrate the wide ranging inferences made possible in this connection when the within condition methodology used in establishing attitude factors in Experiment 24 and the between conditions methodology applied earlier are combined. Since the meanings and importance of any factor are a natural product of the particular scales an experiment employs, comparisons of separate factors in these respects are only valid if the data yielding them are obtained via the same semantic differential questionnaire. Accordingly, the conclusions reported henceforward are based on factor comparisons conducted within the confines of each individual experiment and its various conditions in turn. We draw first, for purposes of illustration, upon Experiment 10.

Between condition comparisons in Experiment 10 indicated that *mediation via TV significantly reduced the differential impact of two contrasting verbal communications*: fewer significant differences were perceived between a positive and a negative political text when they were presented in a TV format than when presented on paper. In the textual (i.e. non-TV) conditions the between condition differences were mixed, the 'positive' version seeming more HUMANE and PLEASANT than the 'negative' version but less SINCERE and, psychologically if not verbally, more NEGATIVE! Within condition examinations of the factors associated with these significantly affected scales reveal the conflicts between scales to have arisen primarily within ratings of the negative text. The ratings of both texts demonstrate a tendency towards attitude conflict, though the factor beset by it in the negative condition is nearly three times more prominent (accounting for 46.3 per cent of the common variance) than that observed in the positive condition (16.6 per cent).

On the former factor, ratings on the POSITIVE/NEGATIVE and SINCERE/INSINCERE scales—known from the raw data to have been relatively favourable—enjoy a high consonant relationship with those on STRONG/WEAK and SHARP/DULL. Since perceived sincerity and strength have been shown by Experiment 24 to be related to the Integrity and Mastery judgement criteria respectively, we may interpret this tendency (see Figure 5.2) as denoting a favourable reaction towards the likely author of the negative text based on attributions concerning his inner qualities. Simultaneously this tendency bears a high inverse relationship with negatively directed scales such as HUMANE/RUTHLESS (Empathy) and MODEST/IMMODEST. The factor as a whole, and thereby 46.3 per cent of the common variance within the subjects' ratings, may thus be interpreted as arising from a tendency to hold the inner and outer qualities of a communicator in conflict: to accord 'ruthless immodesty' distinct respect! In the political compaigning context of this experiment at least, the more RUTHLESS a communicator appears the more STRONG, SHARP, POSITIVE and SINCERE he seems to be rated. (Perceived ruthlessness is also associated with high emotionality.)

As briefly indicated in Chapter 3, this effect replicates several previously reported within the experimental literature. (It embodies the stern 'Your Country Needs You' principle of the First World War recruiting poster.) Osgood et al. (1957, p.121) infer a general tendency for the public to assess politicians in terms of their 'malevolent dynamism' vs. benevolent insipidness'. Blumler (1968) draws a similar conclusion (p.235): 'This hints at . . . a way of perceiving politics itself – as a world in which those leaders who are strong are usually not very amiable and those who try to be nice tend to be weak. It is as if the electors are implying that in politics being dynamic and being decent are mutually exclusive'. The dissonant tendency isolated within our own data above (Experiment 10) most closely fits the description of malevolent dynamism!

The factor analysis of one of the opposing conditions in the same experiment, on the other hand, suggests a prominent tendency towards benevolent insipidness (43.9 per cent): relatively favourable judgements of the performer's personal qualities in the 'positive TV' condition have a high inverse relationship with ratings of his sharpness. Thus the more PLEASANT, HONEST and SINCERE the viewers regard him, the more DULL he appears simultaneously. Judgements of the 'negative TV' condition, incidentally, yield the most substantial attitude factor to emerge in the research as a whole: accounting for 61.4 per cent of the condition's common variance it represents a consonant general assessment, relatively favourable in its orientation. Evidently the combination of two particular presentation elements (a TV format and a negatively worded text) has served to co-ordinate the audience's ratings on several judgement criteria at once with an exceptional degree of conformity. Comparable summative effects between attitude cues have been identified earlier in Chapter 4.

The observation of conflict as well as harmony in a group's attitude patterns can be explained in the theoretical terms introduced in the previous section. These have asserted that consonant attributions by a group of subjects may be predicted as long as the factors on which they are based yield complementary information. On the same basis we may conclude that dissonant attributions are based upon factors yielding conflicting types of information which, across the subject group as a whole, remain unresolved. A favourable impression of a performer's mastery arising from, for instance, the manner in which he is presented by the TV camera, may well be accompanied by equally compelling reservations formed quite independently as to his empathy and poise. As indicated in Chapter 4, such conflicts may arise either from stimulus ambiguity or from audience ambivalence; and they may equally arise within the ratings of an individual subject–as the state of 'cognitive dissonance' ascribed by Festinger (1957) to a logical inconsistency between two simultaneously apprehended pieces of information–or between the ratings of separate subjects due to differences in psychological outlook. (Before the within subject and between subject variances underlying a set of semantic differential data may be separated, refinements in the factor analytic procedure must be made, as Chapter 6 indicates.)

The notion that dissonant attributions may derive from separate sources of information in conflict with one another is conveniently checked in terms of Experiment 9–the only experiment in our series in which, using videotape editing procedures, ambiguous situations were deliberately created. As may be recalled, a single lecture performance was presented in the context of contrasting studio introductions and (televised) audience reactions, both factors being varied simultaneously. Two conditions were created in which the contextual factors reinforced one another: (a) introduction and reactions were both positive; and (b) they were both negative. A further two conditions brought these same factors into conflict with one another: (c) the introduction was positive while the reactions negative; and (d) the reverse of (c).

Basically, *the provision of prior information about the lecturer reduced the effects of televised audience reactions on attitudes towards her.* In addition, the unrotated factor solutions obtained for each condition subsequently reveal the following attitude patterns. Both of the 'harmonious' conditions (a) and (b) yield, by the 'eigenvalue one' criterion, at least four significant judgement factors. The on-balance highly favourable reactions to condition (a) are dominated by a substantial first factor accounting for 56.2 per cent of the common variance in the solution as a whole. No particular loadings predominate over others within the factor and no conflicting loadings are noted; it may thus be interpreted as a general factor denoting the overwhelming tendency within this viewing group towards an uncomplicated assessment of the performer, which the raw data indicate to have been generally favourable in its orientation. The second condition–also unambiguous though on this occasion featurint consistently negative detail about the performer–also yields a first factor identified with general attributions, not surprisingly more negative in character. This factor denotes a far weaker general tendency than was evident in condition (a), in that it accounts for only 26 per cent of the common variance: however, as before, no conflicting loadings are noted upon it.

The two ambiguous conditions, by contrast, each yield factors suggesting powerful conflicts between particular attitude scales. The effect is particularly marked in condition (c), combining positive introduction with negative audience reactions. The judgements in this condition are dominated by a first factor accounting for 38 per cent of the variance and featuring a significant–i.e. loadings greater than plus or minus 0.55–inverse relationship between assessments of the performer's Mastery (e.g. PROFOUND/SHALLOW, STRONG/WEAK) and her Integrity (HONEST/

DISHONEST). An identical relationship is observed in condition (d), albeit denoting conflict between Mastery and Integrity at a secondary level (23.3 per cent) only. Indeed, in each of the three conditions which contains an element or elements of negative detail about the performer, a primary or secondary factor indicates this same tendency to prevail at a relatively high level within the group's attitude structure. Condition (a) alone is free from the effect.

Ideally we should now trace connections between the stimulus qualities brought into conflict in the last two conditions and the two dissonantly related judgement criteria observed simultaneously; and, if it were possible for us to distinguish the criterion underlying the positive ratings in each condition from that underlying the negative ones, this matter would be relatively straightforward. Unfortunately, the factor analytic methods generally applied in the social sciences, and on which the present results are based, do not always disclose the characteristics of a relationship between two variables with adequate precision for this purpose. Thus they do not independently specify whether an inverse effect tends in one particular direction (e.g. towards + Mastery and – Integrity, or vice versa) or, within the subject group at large, *in opposing directions simultaneously*. At times we are able to deduce the direction of a within group tendency from the between group effects associated with it, though we require a technique more finely tuned to the differences between the scores of individual subjects if we are to uncover the multiple origins of an attitude effect with any certainty (see Chapter 6). The shortcoming of standard factor analyses in this connection has been acknowledged in the procedural section of Experiment 19 and by Rummel (1970, section 12.3).

Subsequent experiments may reveal whether, for instance, the studio introduction variable of Experiment 9 consistently affects Mastery attributions, while the televised audience reactions influence perceived Integrity. Numerous similar possibilities may be inferred from the analyses conducted to this point, though many of them must certainly be considered specific to the conditions investigated and with little general application. Accordingly, in the remaining sections of this chapter, we concentrate on the effects which prevail across different conditions thereby suggesting the clearest guidelines for future research and practice.

We classify these effects, from one section to the next, according to three major sources of information evidently used by TV viewers in the attribution process. Disclosed already by the between condition comparisons of Experiments 1 to 17, these are:

1 The TV performance itself.

2 The visual reactions of others apparently viewing the performance.

3 Other detail mediating it to the viewers.

Each of these factors, as we have seen, is independently predicted by Kelley (1967) in terms of attribution theory. A fourth variable recognised by Kelley (p.194)—i.e. the extent to which an individual viewer's responses to media material may vary over time—has not yet been considered by us, though further work in this connection is at the planning stage. Of course, just as each of the above non-verbal factors interacts with each of the others, so they also interact with the host of verbal factors understandably minimised during these experiments just as in other areas of non-verbal communication research. A detailed taxonomy of verbal presentation variables considered worthy of future examination by TV researchers (pacing and rhythm, order and balance of opposing arguments, etc.) is given by Coldevin (1979).

The TV performance

When TV performance characteristics have been varied during the experimental series, verbal content and production stylistics being held constant, it has become evident that certain types of performance conventionally encouraged may have effects quite opposite to those that are intended. In Experiment 17, for example, *'a friendly' TV appeal was found less effective on the donation measure than a 'tense' appeal*, an outcome unlikely to be predicted intuitively, or even by reference to the basic (between condition) attitude effects reported in parallel. However, factor analytic examination of the attitude data reported in Experiment 17 reveals tendencies underlying them that render the behavioural effect more easily understood.

The factors extracted in the 'friendly' condition are unsurprising: the performer's relatively favourable ratings on attitude scales measuring perceived relaxedness, strength, expertise, etc. are found to be consonantly related on a Professional factor accounting for 41.3 per cent of the condition's common variance. In the 'tense' condition, however, the same rating scales were evidently used with more discrimination, being identified with three independent attitude factors and with a consequent 83.9 per cent of the variance in the solution as a whole. While the first of these factors denotes a prominent general tendency associated with, amongst other qualities, perceived strength, the second gives clear evidence of attention to the performer's nervousness and hesitancy, and of an equation between both of these attributes and perceived humanity—a relationship which, according to our original positive/negative coding of the rating scales at least, involves a degree of conflict. The third factor in the 'tense' condition also comprises an inverse relationship—this time between perceived relaxedness and persuasiveness—indicating that the more TENSE the performer appeared to subjects the more PERSUASIVE he was rated, and/or vice versa. Each of the latter two factors thus reveals a conflict between specific qualities and perceived Poise with which the superior behavioural ratings of the 'tense' performance may tentatively be associated.

A very similar effect is noted in the mid shot conditions of Experiment 15: when appearing 'angry' in mid shot an actor was considered more INTERESTING and PROFOUND than when 'friendly'. In the former condition both of these scales are consonantly related to a substantial general factor accounting for 52.9 per cent of the variance; however, in the 'friendly' condition, less favourably viewed, perceived interest value and pleasantness is significantly associated with a high degree of confusingness (15.1 per cent)! Furthermore, *an increase in camera focal length in Experiment 15 was found to enhance the performer's perceived tension even when the (friendly) style of his performance minimised it.* In the 'friendly close up' condition, ratings of greater relaxedness are subsumed within a general factor accounting for 61.2 per cent variance; in the 'friendly mid shot' condition, however, perceived relaxedness is evidently considered synonymous with strength and expertise while negatively related to humanity (26.6 per cent). Thus two of the 'friendly mid shot' factors, accounting for 41.7 (= 15.1 + 26.6) per cent of the variance *in toto*, are characterised by conflicts between professional and personal attributes almost identical to those observed in Experiment 17.

In both Experiments 17 and 15, therefore, factor analysis consolidates the possibility discussed earlier that viewers regard a casual TV manner as to an extent contrived and somehow concealing the truth about a performer. Moreover, Experiment 15 demonstrates the interaction of performance attributions with those arising from an independent variable (camera focal length) mediating it. A relationship between performance style and studio technique–viz. the use of autocue–was also observed in Experiment 13, in which the latter's impact upon the degree of camera eye contact *affected a newsreader's perceived caution, precision, tension and directness*. Factor analyses of the three conditions in this experiment reveal prominent inverse relationships in each. When giving the camera lens minimal eye contact the newsreader was judged primarily in terms of a high degree of tension (38.7 per cent); tension in this condition is significantly related to caution though also to weakness. When his attention is shared between the desk script and camera lens more equally, his perceived caution again connotes weakness in the factor solution though also more favourable qualities such as expertise and humanity (25.3 per cent). Ratings of caution/rashness, it may be recalled, were significantly higher in the intermediate condition than in the third condition, in which the camera lens and autocue device were fixated a great deal as per conventional TV newsreading practice; the newsreader's relative rashness, as perceived in the third condition, is significantly related to forcefulness

(15.2 per cent), suggesting that in Experiment 13 at least rashness may have been regarded as a positive rather than—as commonly supposed—an adverse quality. Ratings of this condition are dominated, however, by a prime tendency (33.3 per cent) to associate significantly higher degrees of perceived relaxedness and directness with reliability and yet intolerance, a further sign of suspicion towards the polished TV performance reaffirming the need for either moderation in autocue usage or for training in the art of using it to the least ambiguous effect.

Other effects possibly due to autocue usage have occurred in Experiment 3, in which, *when seen to address the camera directly, a performer was considered less reliable and expert than when seen in profile*. On this occasion the performance itself was not varied, though as far as the viewers were concerned different features of it may certainly have been intensified in each of the conditions. Thus it was earlier hypothesised that the differences between conditions may have derived from the variable degree of camera eye contact provided by them. An alternative hypothesis suggested that the two conditions are conventionally associated with different professional functions (newsreader, expert, etc.). In fact, Experiment 14 has indicated that both hypotheses may be tenable and mutually compatible, for *when varied simultaneously camera angle and eye contact had selective effects on performer ratings*. In this experiment, as in Experiment 3, the performer was rated more EXPERT in profile than when addressing the camera directly: when his eyes were obscured electronically, however, he was considered more RELAXED than when viewed normally. Thus camera angle and gaze characteristics appear to have cued attributions regarding expertise and tension quite independently. The relationship between profile performances and higher perceived expertise has also been found in Experiments 6, 7 and 8 (i.e. on five separate occasions in all). On no occasion has the effect failed to emerge, though in the multivariate experiments it often did so in interaction with other cues.

The factor analyses of Experiments 3, 6, 7, 8 and 14 consistently equate judgements of expertise in the 'profile' conditions with a prominent tendency, free from conflict, towards judgements of the performers' inner qualities of Mastery and/or Integrity. In each (except for Experiment 3) judgements of the 'direct' condition are by comparison dominated by a tendency towards conflict between professional and personal judgements. Yet again it emerges that viewers cannot always reconcile these two criteria: and in the more artificial, direct to camera contexts, the problem appears to be accentuated.

Conflicting judgements of professional and personal qualities are also observed in the three conditions of Experiment 11, in which *a performer's visual and vocal qualities were rated differentially*. This experiment also employed a 'direct' address, maintaining it throughout both visual conditions. The between conditions results have indicated that separate performance channels can exert independent attitude effects; and, as we see in the next two sections concerning the effects of production variables *per se*, particular mediating strategies may be used to direct audience attention between specific aspects of a performance quite deliberately.

Visual commentary

Two experiments have demonstrated the effects upon a performer's impact of televised audience reactions: i.e. the influence of other persons (cf. Kelley, 1967). When interviewer reaction shots are presented, the viewers are evidently highly sensitive to any clues, witting or otherwise, that they may provide as to the ability and nature of the interviewee. In Experiment 8, for instance, an impression of relative tension in a profile performance (condition B) was diminished by the insertion of interviewer reactions in condition D. Judgements on the RELAXED/TENSE scale in both conditions are identified with a consonant general factor, more prominent in B (47.1 per cent) than in D (26 per cent). In the latter condition tension ratings were presumably less systematic than in the former because the interviewer reactions within it provided a 'visual commentary' rendering the condition more ambiguous to the audience as a whole.

It is naturally likely that any effect such as this will vary directionally according to the precise nature of the reaction shots perceived. When in Experiment 4 televised reaction shots were indeed varied, their insertion into a televised lecture had contrasting effects as expected. Positive reaction shots yielded relatively favourable assessments of the lecturer associated on a consonant general factor (41.2 per cent); a secondary, specific factor (21.1 per cent) is identified with favourable assessments of her Mastery. When punctuated by negative reaction shots, the same performance was again judged on a general basis though one naturally more negative in its orientation. The general factor in question (37.7 per cent) is identified with the same scales (denoting Mastery and Empathy) as affected in the other condition: in addition, however, there exists in the 'negative' condition a clear secondary tendency (23.5 per cent) for judgements based on these two criteria to conflict. Relatively low ratings on the profundity, expertise and strength scales are related to high ratings of pleasantness. This effect suggests that viewers were in little doubt that the performer was less 'masterful' in this condition, though were prepared to sympathise with her personally for this very reason. It is a version of the delightfully named 'benevolent insipidness' effect alluded to earlier (Osgood et al., 1957, p.122). Once again the factor analytic method has served to extend an earlier conclusion, this time concerning the *effects of audience reaction shots on the impact of the lecturer's message as well as more personal qualities.*

The edited cutaway technique conventionally used in TV news broadcasting to create reaction shots may, as seen in Chapter 2, influence attributions in its own right. The reactions, being faked, appear more tense than in a normal conversational situation, and particular juxtapositions of material created by the film editor may also affect a performer's impact in quite surprising ways. In Experiment 5 the technique was turned to good effect, partly owing to the care and skill applied in the editing process: *the re-recording and edited insertion of the interviewer's role in a discussion (a) increased his perceived tension and sincerity as well as his apparent intelligibility: and (b) substantially enhanced the general impact of the interviewee.* When the discussion was presented via two cameras, factor analysis shows the interviewer to have been judged on three independent bases. One represents a tendency to identify his relatively high relaxedness in this condition with perceived unreliability and inferiority (Poise vs. Integrity: 2.5 per cent). Tendencies towards relatively unfavourable judgements of general qualities (30.2 per cent) and Mastery (14.5 per cent) are also noted. In the edited condition, however, general assessments were improved, and their overall standing within the factor solution increases to 46.6 per cent–i.e. they are more systematic–the other judgements significantly affected by the cutaway technique are co-ordinated within a further Poise vs. Integrity pattern (30.2 per cent).

The side effects of the interviewer cutaway shots on perceptions of the interviewee were quite dramatic. In the unedited condition he was judged significantly less favourably on 8/16 scales. Six of these are identified with a very prominent negative evaluation accounting for 50.9 per cent variance; the remaining two scales are identified with, once again, a 'benevolent insipidness' effect–i.e. HONEST and HUMANE but WEAK and INSIGNIFICANT (25.6 per cent)! In the edited condition judgements of the interviewee yield a relatively positive general factor (35.7 per cent) and a comparable Mastery factor (28.5 per cent). While the Mastery criterion is more prominent in the viewers' assessments in this condition than in the other, it may be observed that the general factor is less so. Though we have not explored the question directly within this research, it appears likely that a methodology for disentangling the image factors causing covariance of viewers' attributions may be developed on the basis of judgement tendency estimates yielded in this manner.

Less skilfully simulated reactions on the interviewer's part combined with hasty against the clock editing might certainly render the effects of edited cutaway shots less beneficial than they have appeared here. Although certain effects of image variation are clearly quite consistent, as will be emphasised in Chapter 7, others such as these are a good deal more fickle, deriving from perhaps ambiguous perceptual cues which take on different connotations according to context. The effects of general peripheral detail in a TV production are evidently of this type.

Peripheral detail

Having illustrated the manner in which they are drawn we will summarise the remaining factor analytic conclusions in less detail. Certain of them are in any case quite tentative and must be tested in a further range of conditions if they are to be validated. In the experiments examining the effects of general visual detail upon performance impact, however, one maxim seems clear: that in ambiguous viewing situations they are powerful and amenable to a high degree of control by production staff for particular purposes. If serving a simple arousal function, stimulating the viewers' attention in otherwise boring or confusing conditions, peripheral details have strong and usually quite predictable effects. Thus in Experiment 2 *the 'keyed' insertion of a picture background increased a newsreader's perceived credibility rather than his simple interest value.* In the 'plain' condition, conflict is noted between relatively positive judgements on the Mastery criterion and others denoting an Integrity assessment ('malevolent dynamism'?). This conflict is not apparent when the picture background is perceived: impressions of the performer's Integrity in this condition are enhanced on several scales, and positive judgements of his Mastery are formed independently. A very similar effect is noted in Experiment 6 on the addition of 'notes long shots' (condition C) to an otherwise unrelieved 'direct close up' (A): a tendency in the latter context to equate perceived relaxedness with weakness–and/or tension with strength–is overwhelmed, and a consonant positive assessment of the performer's Mastery takes its place.

That viewers in Experiment 6 were indeed highly susceptible to the simple arousal function of visual detail was also indicated by the between condition contrasts of C vs. E, and D vs. F. However, in contexts in which the basic performance assessment is generally more favourable than in Experiment 6–and quite unambiguously so–added visual detail may serve to irritate rather than to arouse and to create judgement conflict where none existed previously. Thus when, for the first time in the current series, a professional newsreader was viewed in Experiment 12, judgements of his performance set against a plain background were dominated by a positive general tendency, *an impact which the keyed insertion of a relevant picture background reduced.* This effect is accompanied by a significant reduction in his perceived Empathy and by conflict in the 'picture' condition regarding his personal qualities generally. Evidently, as concluded in Chapter 3, an already favourable impact may be hampered by an over-zealous attempt to enhance it further via image manipulation: 'the subject's attention is diverted away from an important causal factor' (Kelley, 1967, p.231).

The most predictable function of added detail is therefore one of increasing general viewing attention in situations most starved of stimulus. Within a single highly detailed image, however, numerous probable sources of meaning may be identified. If a single unit of meaning is varied in isolation—the 'smallest signifying unit' or 'seme' (Fiske and Hartley, (1978)—it may exert logical effects quite opposed to those predicted on an arousal basis. Thus in Experiment 1 *evidence of a lecturer's use of notes diminished viewers' perceptions of his fairness and straightforwardness*, an effect mirrored by the substantially greater prominence of Mastery assessments in the 'notes' condition generally. Via tight control over the 'notes' detail in a suitably ambiguous performance situation, the very subtle connotations of a lecturer's reliance upon them were allowed to exert an effect in their own right. Whether in a particular context individual cues serve a general arousal function or one more subtly based is not always easy to discern: '. . . signs can belong to more than one aesthetic code, and so codes can overlap and interrelate in a network of signification . . .' (Fiske and Hartley, p.64). When sufficiently powerful in their own right, image cues may have quite predictable positive or negative attitude effects, as exemplified by those of the televised audience reactions in Experiment 4; and when different cues summate—as in the children's data of Experiment 7 (condition E)—they may produce a predictable behavioural effect also (see Experiment 21). But, of course, different subjects do not necessarily perceive the same cues in a piece of TV material, or (if they do) are not necessarily inclined to use the same criteria in responding to them. The problems of prediction and control over image effects that stem from audience differences specifically are considered in the next chapter; the immensity of this question, however, may first of all be anticipated in terms of our 'BVT' data.

Adults vs. children

In checking for differences between the effects of image variation upon adults and children (see Experiments 6 to 8) we were quite unprepared for the wide range of differences that emerged. Whereas ratings by the adult subjects in Experiments 6 and 8 were influenced by selected image cues only, and on a relatively narrow range of judgement scales, those of the children in Experiment 7 suggested a sensitivity to each one of the cues experimentally manipulated and differential effects on all but a very few of the scales administered to them. Thus *added location detail, evidence of a speaker's use of notes, interviewer reactions and variable camera angle had often ambiguous effects on his perceived qualities.* The subtelty of effects in this experiment is only fully glimpsed followint the factor analyses.

When location detail is included within the presentations (in conditions E and F) the children are highly attuned to it, and the first factor in each of these conditions denotes a powerful (positive) general assessment. The children's attributions towards the performer are formed on no simplistic basis, however, for individual scales are interrelated on a wide range of factors indicating the usage within each viewing group of a variety of judgement criteria. Thus a tendency is noted in one condition (C) for subjects to rate the performer in terms of low professionalism, and to equate the latter with a low level of rehearsal (i.e. SPONTANEOUS); in B,Integrity is equated with rehearsal similarity; in A, a clear Mastery factor is noted; and so on. Other prominent factors underlying the children's data are identified with professional and personal judgements respectively, and–particularly in the profile conditions–with Empathy and Integrity factors.

Both the 'notes' and 'interviewer reaction' variables have the types of effect upon children that demand a more subtle level of interpretation than is afforded by the simple arousal notion. As in Experiment 1, the addition of 'notes' detail (condition C) significantly diminished the performer's impact though on the basis of independent general, Professionalism and Mastery criteria. Prior to the notes insertion in the 'direct' condition (A) the children reveal a tendency towards a 'malevolent dynamism' judgement–viz. conflicting judgements of Mastery and Integrity. This effect resembles the other dissonant effects observed in 'direct' conditions elsewhere in the data. The addition of interviewer reaction shots to the basic profile condition (D) has a dual effect for these subjects, traceable by factor analysis to independent judgements of the performer's inner and outer qualities. The reaction shots enhance judgements on the former criterion though diminish them on the latter. The comparison of basic direct and profile conditions (A vs. B) yields a similarly ambiguous effect: in the latter context the performer's perceived professionalism is significantly reduced though his personal attributes are enhanced.

Considered as a whole, the analysis of children's responses to image variations suggests quite categorically that their capacities to draw inferences from non-verbal images on highly subtle bases are greatly underestimated. Yet they may well lack the critical skill necessary to evaluate their attributions, and to recognise that those based on mediating factors alone probably have little or no relevance to questions of a performer's personal or professional characteristics. The adult is likely to have learned this lesson, or at least partially so, from his experiences of human nature in a wider range of situations (see Experiment 8). For the child, however,

TV may actually be the only medium by which the wider vagaries of human nature have been mirrored, and the effects of mediating technique upon his attributions towards the persons he perceives by this means will therefore be more substantial than at a later date. Individual effects upon children of the type we have summarised should clearly be subjected to much further scrutiny; for unless we can understand and predict the judgement criteria applied by children in complex perceptual situations, we shall be unable to help them to resolve the conflicts they evidently experience as a result, or to overcome their potential vulnerability to persuasive influences of an abusive kind.

The statistical methodologies applied during the early part of this research, by which image effects upon subject groups were examined according to the classical criteria for between-condition variation, do not actually assist us to predict and unravel audience differences in the most efficient manner. Our insights in this respect are partly restricted by the broad functional distinctions we observe when selecting our experimental subjects for testing in particular conditions ('student', 'adolescent male', 'middle class female Caucasian', etc.); and since the normal methods identify a broad consensus of tendencies within the subjects' scores, numerous important though subtle experimental effects doubtless pass unnoticed by them. In the present research context factor analysis has been applied in order to throw light, diagnostically, upon effects not disclosed by the between-condition methods. Yet conventional factor analysis also has its blind spots and raises questions regarding the data that it is unable to answer.

For example: do the individual tendencies to judge TV material on different criteria co-exist in each of the audience's members? or do they reflect independent sub-groups within the audience, each viewing the material in a characteristic way? Likewise: are the conflicts between judgement criteria (e.g. professional vs. personal) experienced by individual viewers? or do these too represent conflicts in viewing style between separate audience factions? In Chapter 6, we refine our analytic methods in order to give answers. Until able to do so, it is likely that the logical effects of, for instance, TV background music (Experiment 16) will remain unclear. Musicality being a more specialised gift than general visual ability, it is possible that the effects of music on different individuals vary and contrast to an unusually high degree, requiring an appropriately more sensitive style of analysis. The only general effect evident in Experiment 16 was that *the addition of solemn soundtract music to the beginning of a film documentary extract reduced the subsequent speaker's perceived reliability.* Certainly a clear tendency is noted in the conditions featuring 'joyful' music, towards relatively consonant and favourable general judgements, while in both of the conditions featuring 'solemn' music marked dissonant judgement tendencies are noted. We can only conjecture on this basis that the solemn music jarred within the groups' responses with the highly pleasant and professional performance presented, and was particularly effective in reducing aspects of its impact when preceding it since creating a preparatory 'set' (cf. Postman and Egan, 1949).

On the basis of the analytic methods used hitherto, however, we may certainly draw several more confident conclusions than this in general. Throughout the above experiments we have observed the duplication of a number of image effects interpreted, indeed often predicted, according to general social psychological theory. We have also observed the repeated emergence in the data of effects identified with particular rating scales. Thus judgements on the EXPERT/INEXPERT scale prove a quite consistent index of differential image effects, possibly due to the scale's dual association with both Mastery and Integrity: having two meanings in general application (see Experiment 24) perceived expertise has in effect a double chance of being affected by image variations. Indeed both of the scales found to have a dual meaning, or 'variable complexity of two', in our own research coincide with those factors singled out by Kelley (1967, p.204) as most commonly found relevant to the acceptance of communicated information generally (viz. 'expertness' and 'trustworthiness'). It is clear that only by attention to the levels of complexity at which attitudes are formed is the highly variable nature of message acceptance to be adequately explained.

The principal outcome of our own emphasis upon attitude complexity has been the evidence that image variations frequently lead to types of judgement conflict capable of quite specific interpretations (e.g. concerning inner attributes vs. outer, professional qualities vs. personal, and other even more specific criteria). Many of these effects, in running quite counter to conventional production intuition, give food for thought that the TV practitioner may well care to consider. If a person seen on TV is expected by his audience to fulfil the function of a TV professional, for example, a high degree of poise will probably be expected of him—but otherwise not. The TV professional on the other hand is not necessarily expected to have intellectual mastery over the information he conveys (see Experiments 1, 2 and 13). Nor is the professional of any kind bound to demonstrate a high degree of empathy in his public dealings. The frequently observed types of attitude conflict that have led us to these conclusions reflect a general suspicion on the viewers' part, either individually or collectively, regarding many styles of TV performance commonly encountered. The 'benevolent insipidness' attribution principle, for instance, suggests that 'the more charming the performance the less it is to be respected'! The 'malevolently dynamic' performer on the other hand is one who fits such maxims as, for example, 'a stern performance radiates high integrity', and 'a real professional does not need to curry his audience's favour'. The evidence of such viewing maxims as these throughout our data, regardless of viewing type, certainly suggests that the TV audience in general is a far more sophisticated judge of the persons presented to it than is commonly imagined.

We have not been able to establish exclusive connections between media practices and particular attitude patterns, though future studies in which attitude measures are carefully balanced according to the specific judgement criteria uncovered by Experiment 24 may certainly be conducted for this purpose. The evidence to this point indicates that particular image variables may affect reactions based on a number of criteria at once, and that specific judgement criteria may be simultaneously affected by several image cues. Work by Sells (1979), however, has certainly suggested that scales identified in the present study with Empathy and Integrity are particularly vulnerable to a direct/profile variation: in demonstrating the viability of a signal detection approach to the study of image effects (cf. Baggaley and Duck, 1976, p.155) he has established that ratings on both of these factors are more consistently affected than those on the other two, and has obtained results further supporting the general preference for performers seen in profile. Similarly, Sullivan et al. (1979) have demonstrated the predominance of Empathy and Integrity effects in an experiment comparing perceptions of a performer seen on television and face to face: though the performances in both conditions were identical, the ratings in the latter were significantly more favourable in these personal terms than those in the former!

All work of this type serves to clarify the effects of media communication in general and specific contexts. But, until the variety of effects upon individual audience members can be predicted with the same accuracy as the broader effects upon audiences as a whole, practical applications of the work will be limited. The insensitivity of conventional variance and factor analyses to uncontrolled differences between individual subjects represents their main inferential weakness in the present context; and in the next chapter an attempt is made to deal with the problem.

6 Predicting audience differences

Throughout the earlier chapters we have referred exclusively to group-based evidence for TV image effects: i.e. to the effects of image variables that prove most uniform when the data of individual subjects are combined. Thus we refer to the effects that emerge as a general consensus of tendencies within subject groupings,and to the ways in which these general tendencies differ as a result of separate treatments. By employing within group as well as between group statistics we are able to show that an audience's response tendencies may conflict with as well as supplement one another, and to recognise that image and attitudinal variables fall into distinct categories capable of quite complex interactions. On this basis we gain several insights into the conscious and unconscious effects of TV production that may prove pragmatically useful in different applied contexts.

Yet there are still many effects to which we are unable to attribute firm causes. The tendencies towards conflict within a group's attitude ratings in particular suggest the invisible agency of a range of psychological factors which the standard experimental methods do not reveal, and yet whose origins we must certainly identify if the variety of image effects is to be fully understood. At this point, however, we are threatened with an impasse. For, as indicated in the previous section, the effects we have identified so far may arise either within the data of individual subjects or between the psychologically differing members of the population at large; and our ability to tell whether one or both of these possibilities applies to a given effect is dependent on the precision with which we are able to pinpoint the relevant individual differences between our subjects on the basis of objective criteria. But as long as the psychological factors underlying the effects we observe are themselves unknown, how are the criteria for classifying our subjects in this connection to be recognised? And how, without adequate predictive criteria, are the psychological variables relevant to our inquiry ever to be determined?

When the effects of an experimental treatment are reflected by measurements on a relatively simple, one dimensional basis, e.g. level of performance in a memory task, a straightforward solution to this problem is available. Subjects' scores on the task may be used as the objective criterion for assigning individuals to separate sub-groups for further examination. Comparisons between sub-groups, and interactions between their separate scores and further experimental treatments, may then gradually reveal a range of psychological determinants previously unsuspected. This course of action was recommended by Cronbach (1957, p.681) 'Applied psychologists should deal with treatments and persons simultaneously. Treatments are characterised by many dimensions; so are persons . . . (W)e should seek out the aptitudes which correspond to (interact with) modifiable aspects of the treatment.' The need to examine subject-treatment interactions in the educational media context was discussed by Baggaley (1973) and a range of methodological approaches to the question of 'aptitude treatment interactions' specifically has been chronicled by Clark and Snow (1975), Cronbach and Snow (1977) and by Clark (1978).

When attitudes are measured, however, as in the present context, measurement is patently a multi-dimensional question, and the role of particular dimensions in different conditions is more difficult to predict. In this as in any relatively open-ended psychological situation, the criteria for between subject differentiation used in the classification of aptitudes, as above, are plainly inadequate. In the absence of other criteria, however, the only available recourse is to intuition and to functional subdivisions of the population which tend to be sweeping, speculative, and often only indirectly related to the question being investigated. Suspecting for example that age, intelligence, and/or experience affect our results in particular situations, we may divide our subjects into *a priori* groupings of adults versus children (see Experiments 6, 7 and 8) and hope that the differences between their scores provide adequate vindication. If they do, however, we will be no closer to predicting whether age, intelligence or experience bear upon other situations equally; or even whether the three variables are equally important in the situation actually tested; for their interrelationships are inferred on general evidence rather than defined with respect to the particular parameters at hand. Indeed we have no guarantee that in particular subjects the individual variables are actually related at all, and little opportunity to prevent the experimental outcome from being hampered by perverse interactions between them due to extraneous factors which they variously reflect.

Thus, if a member of an adult subject grouping tends to view TV extracts in the manner of a child or vice versa, his responses will serve to diminish the differences observed when adult and child subject groupings are compared. For reasons quite unconnected with the validity of the underlying hypothesis, the experiment may therefore fail. The inadequate grouping criterion alone will be responsible for this, having failed to allow that a childlike responsiveness is effectively measured less in simple chronological terms than in relation to the perceptual and judgemental skills with which different age groupings are generally associated. Since all skills are effectively modified by the treatments they receive, moreover, it becomes apparent that *a priori* criteria for subject classification can only be fully reliable when the treatment effects are totally predictable: and this, if the experiment is worth conducting, is highly unlikely.

The problems of *a priori* sampling in these respects beset all areas of psychological enquiry in which the role of individual differences comes to be recognised. Within media research, as indicated in Chapter 1, current speculations seem likely to generate more intensive investigations of individual reactions to contrasting treatment styles than have been fashionable hitherto. For instance, Eysenck and Nias (1978) suggest the need to seek for connections between media effects, extraversion and neuroticism. Yet (Baggaley, 1973, p.162) these broad personality typings are themselves merely generalised reflections of the perceptual and judgemental skills with which mediating factors may interact more directly; and attempts to rationalise media effects on *a priori* personality bases are therefore likely to be fraught with the same problems as discussed above. In view of current developments, therefore, it seems important that objective criteria for subject differentiation and groupings should now be established; and in contexts where perceptual and/or judgemental tendencies are directly measured it is desirable that the between subject differences dictating these criteria should be inferred from the experimental data themselves.

Patterns vs. magnitude

In order to establish the factors actually differentiating subjects in their responses to experimental treatment, a criterion test is required capable of identifying whether an individual subject—according to the particular measures used—truly represents the group to which he is assumed to belong. Predictions from the data may be weighted accordingly. Following the collection of, for example, audience responses to media material, the major directions taken by them need to be determined so that subjects with exceptional scores may be set to one side. As a means of examining the cohesiveness of subjects in response to experimental conditions such a test would have immediate uses, identifying the main sub-groups of a given mass for further inspection and the relationships of each subject to them. In separating different factions within a mass one from the other the technique would stress the similarities and differences arising between subjects, and would thus distinguish them from the effects due to psychological consistencies and inconsistencies within the individual. On this basis it would provide a valuable introduction to the analysis of social and psychological conflict, and a check on many of the assumptions underlying the study of human tendencies at massed levels.

The question is specifically one of between-subject variance, and one to which many social science statistics do not apply themselves. While variance analyses can compare the variances arising both within and between sets of data simultaneously, conventional correlation methods tend to emphasise the variance within sets of data, erasing that which separates them by reducing both sets to standard scores. Thus product moment, the correlation technique most frequently used in the social sciences, estimates the similarity of two or more sets of data in terms of the peaks and troughs within them rather than of the distance between them. It deals in the data characteristic which Rummel (1970, section 12.3) has classified as 'shape' or 'pattern', overlooking that which he terms 'magnitude' (Figure 6.1). Of course, analyses via factoring techniques of the patterns underlying multiple data relationships merely reflect the information provided by the correlation coefficients on which they operate; and, using conventional techniques, detailed analysis of the factors underlying magnitude differences between data sets is thus impossible.

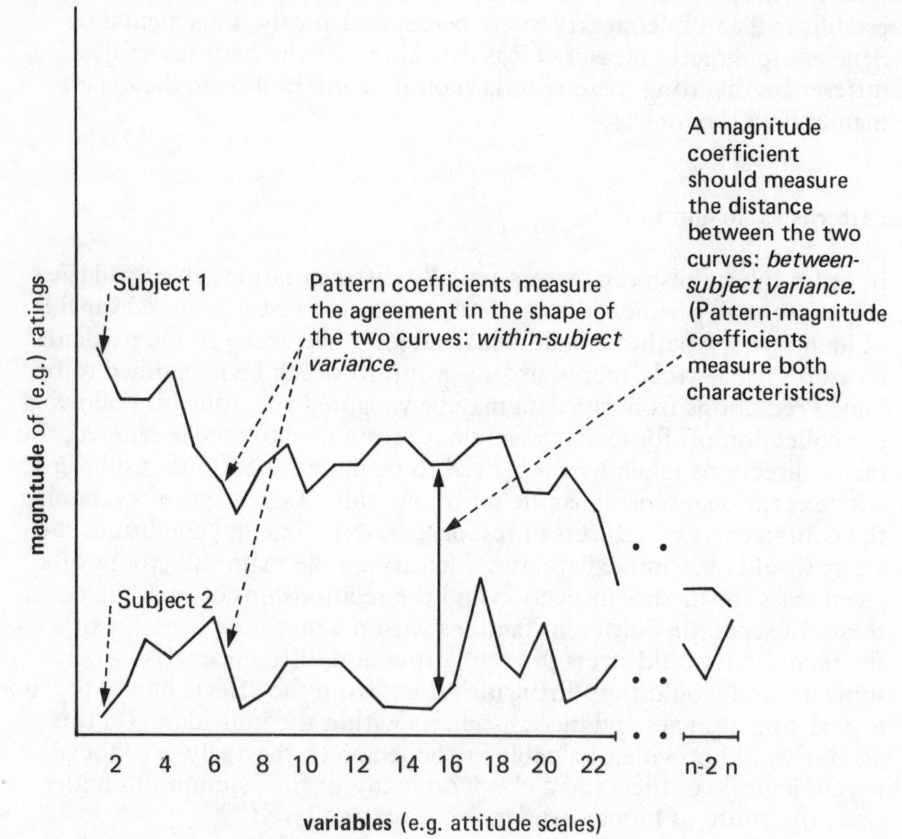

(After Rummel, 1970, p.300)

Figure 6.1 Pattern vs. magnitude characteristics in correlation of subjects' scores

The shortcomings of conventional correlation measures in relation to attitude measurement specifically were recognised in the classical exposition of the semantic differential technique used in the present research:

> (T)he correlation coefficient does not take into account the absolute differences between the means of two tests; perfect reliability, r of 1.00, can occur when an absolute difference . . . exists between test and retest measurements such that not a single score is reproduced and, on the other hand, reliability can be indeterminate . . . when every subject gives exactly the same score on retest as he did on test (Osgood, et al., 1957, p.127).

Naturally, statistical techniques find or fall from favour conventionally according to the styles of research question generally asked; and the common use of the product moment correlation coefficient reflects the common interest of social scientists in the relationships between specific variables measured across n subjects, rather than between specific subjects scoring on n variables. When the relationship of discrete variables is the object of attention, magnitude differences between the n scores on each are usually artefactual and can be excluded from the analysis. But when, as in the present case, relationships between subjects on specific variables are of interest, the absolute difference between their scores is central to the analysis and a coefficient other than the product moment must be used.

Unfortunately, as Rummel (1970, p.297) writes, 'most computer programs of which I am aware compute the product moment matrix transformation as the only correlation option prior to factor analysis', and, as far as the present writer is aware, the observation still holds. Various alternatives to product moment correlation are available, notably 'pattern magnitude' techniques which compute the between subject and within subject variances as in variance analysis (Cattell, 1949; Muldoon and Ray, 1958; Kendall and Stuart, 1961). However, no technique appears to be in current use taking the between subject variance exclusively into account as product moment correlation does the within subject variance. The next section therefore presents a coefficient which expresses the correlation between n scores on paired data axes in this manner; and it uses the coefficients thus yielded in an isolation of the main sub-groups (or factions) underlying N subjects' scores obtained *en masse.* Since a variant of factor analytic technique is utilised, the procedures are referred to collectively as 'faction analysis'.

Faction analysis

The methodology involved in faction analysis is illustrated via a set of artificial data essentially the same as those we have obtained by the semantic differential technique in the earlier experiments (see Table 6.1). Ten subjects (columns) are regarded as yielding ratings of a set piece of TV material on eight bipolar seven-point scales (rows). As previously, high scores on each scale represent unfavourable ratings; however, for ease of illustration, within subject variance in the data is minimised. The methodology's usage is further demonstrated later in the chapter in relation to authentic data and its relevance indicated to other areas of social scientific research in which data are obtained on an interval scale, and in which the sub-groupings of a subject mass on individual measures are of interest.

Table 6.1

Artificial attitude ratings for faction analysis

Scale	*Subjects*									
	A	B	C	D	E	F	G	H	I	J
a	1	1	5	5	5	5	7	4	4	4
b	1	1	5	5	5	5	7	4	4	4
c	1	1	5	5	5	5	7	4	4	4
d	1	1	5	5	5	5	7	4	4	4
e	1	1	5	5	5	5	7	4	4	4
f	1	1	5	5	5	5	7	4	4	4
g	1	1	5	5	5	5	7	4	4	4
h	1	1	5	5	5	5	7	4	4	4

(NB Lower scores denote more positive ratings)

Between subject correlation

1 Each set of ratings (one for each subject) is correlated with each other set in the following manner. Where x_{ij} and x_{ik} are the scores of subjects j and k on attitude variable i, and where n, the total number of variables in the set A, = 8, the between subject variance of the two scores (m) is given by Kendall and Stuart (1961, vol.2, section 26.25) as the total squared deviation from their mean:

$$m = (x_{ij} - \bar{X}_i)^2 + (x_{ik} - \bar{X}_i)^2 ,$$

where $\bar{X}_i = (x_{ij} + x_{ik}) /2$. The total between subject variance of scores (M), summed across all n variables in each subject pair, is given as:

$$M = \frac{\sum_{i=1}^{n} \left[(x_{ij} - \bar{X}_i)^2 + (x_{ik} - \bar{X}_i)^2 \right]}{n} . \qquad (1)$$

If each case has the same two values on the two variables, $M = 0$ (Rummel, 1970, p.302).

2 When two columns of n scores are identical and $M = 0$ ($= M_{min}$), their between subject correlation coefficient (C) should logically $= +1$. When their overall magnitude variance is as great as a given set of interval scales will allow it to be, C should logically $= -1$. When the overall magnitude variance is half that which the set of scales permits, C should $= 0$.

For a y-point rating scale, the maximum between subject variance (M_{max}) is calculated by setting x_{ij} in equation (1) at y, and x_{ik} at unity. For a seven-point rating scale as in the present context, $M_{max} = 18$.

The median between subject variance of scores on a y-point rating scale (M_{mid}) is calculated by setting x_{ij} in the same equation at y, and x_{ik} at the scale's midpoint value

$$(x_{ik} = y - \frac{(y - 1)}{2}) .$$

For a seven-point rating scale, $M_{mid} = 4.5$.

3 As the between subject similarity of the two groups of scores decreases, M increases exponentially; and, in the derivation of correlation coefficient C as a linear function of M with the above properties, M is subjected to square root transformation. The difference between the root transformations of M_{max} and M_{mid} is now equal to that between those of M_{mid} and M_{min} :

$$(\sqrt{M_{max}} - \sqrt{M_{mid}}) = (\sqrt{M_{mid}} - \sqrt{M_{min}}) ,$$

and $\sqrt{M_{max}}$ is twice the value of $\sqrt{M_{mid}}$:

$$\frac{2\sqrt{M_{mid}}}{\sqrt{M_{max}}} = 1 .$$

When the observed between subject variance (M_{obs}) is minimal ($\sqrt{M_{obs}} = 0$),

$$2(\sqrt{M_{mid}} - \sqrt{M_{obs}}) = \sqrt{M_{max}},$$

and the coefficient C is calculated as the ratio of the two halves of this equation, thus:

$$C = \frac{2(\sqrt{M_{mid}} - \sqrt{M_{obs}})}{\sqrt{M_{max}}} = 1. \qquad (2)$$

Thus derived, C embodies each of the properties required above for the extremes of between subject variance observed. It may be computed for an observed value of M according to the relationship plotted in Figure 6.2.

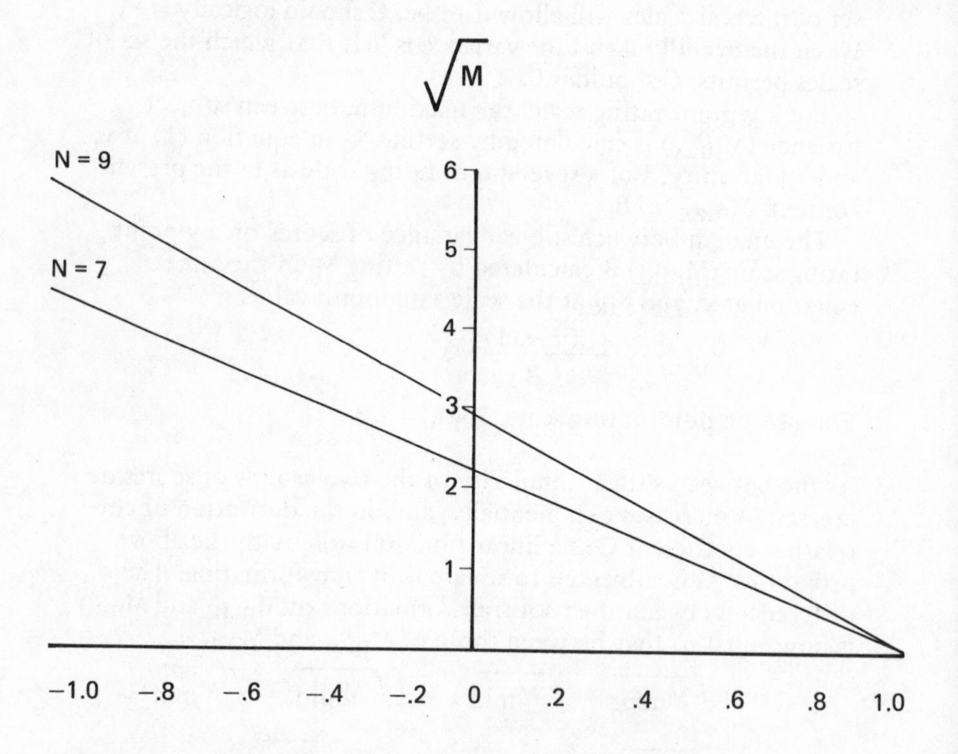

(NB Relationships plotted for 7- and 9-point bipolar scales)

Figure 6.2 Between subject coefficient (C) vs. variance measured (M)

The subject by subject matrix

In a study of the factions within a population of N subjects, the coefficients expressing similarity between the n scores of each subject pair are computed according to formula (2) above and tabulated in an N x N matrix with communalities set at unity. The subject by subject matrix computed from the test data of Table 6.1 is presented in Table 6.2.

Table 6.2

Between subject coefficients computed from Table 6.1

Subjects	A	B	C	D	E	F	G	H	I	J
A	1.00	1.00	-0.33	-0.33	-0.33	-0.33	-1.00	0.00	0.00	0.00
B	1.00	1.00	-0.33	-0.33	-0.33	-0.33	-1.00	0.00	0.00	0.00
C	-0.33	-0.33	1.00	1.00	1.00	1.00	0.33	0.67	0.67	0.67
D	-0.33	-0.33	1.00	1.00	1.00	1.00	0.33	0.67	0.67	0.67
E	-0.33	-0.33	1.00	1.00	1.00	1.00	0.33	0.67	0.67	0.67
F	-0.33	-0.33	1.00	1.00	1.00	1.00	0.33	0.67	0.67	0.67
G	-1.00	-1.00	0.33	0.33	0.33	0.33	1.00	0.00	0.00	0.00
H	0.00	0.00	0.67	0.67	0.67	0.67	0.00	1.00	1.00	1.00
I	0.00	0.00	0.67	0.67	0.67	0.67	0.00	1.00	1.00	1.00
J	0.00	0.00	0.67	0.67	0.67	0.67	0.00	1.00	1.00	1.00

Extraction of unrotated factors

To determine whether the population under study is cohesive in its responses or in fact composed of contrasting sub-groups, the above matrix may now be submitted to a factoring technique. (NB At this stage of the analysis individual factors denoting degrees of 'extremism' or 'moderation' in the data are extracted, and not the distinct sub-groups to be known in due course as factions.)

Though factor analysis of between subject relationships (Q-technique) is less common today than that of correlations between experimental variables (R-technique), it has a longstanding history (Burt, 1937; Stephenson, 1953; Cattell, (1966).As other types of factor analysis (see Chapter 4), it identifies the extent to which the variance in a set of data is ordered systematically and the hypothetical factors on which it is based. Subjects A and B in Table 6.1, in having identical scores and a between subject coefficient of +1, will be expected, on Q-factoring, to relate to any one of the factors extracted to an identical degree. Other subjects may relate to the same factor to a greater or lesser extent, though subjects A and B will be distinguished among them by the similarity of their relationships to each factor in the general pattern.

The main factors extracted by a principal factoring analysis (Nie, Hull et al., 1975, section 24.2.1) from the test data of Table 6.1 are presented in the unrotated factor matrix of Table 6.3 (a). (NB In normal applications of faction analysis it is desirable for at least 20 subjects per group to be tested so that their sub-groupings might be of a size sufficient to allow for further analysis. Use of the non-iterative method of principal factoring is recommended and used here accordingly.)

Table 6.3

Q-factor solutions computed from Table 6.2

Subject	*(a) Unrotated factors*			*(b) Oblique factors*		
	I	II	III	I	II	III
A	-.402	(.897)	.183	.00	.00	(1.00)
B	-.402	(.897)	.183	.00	.00	(1.00)
C	(.955)	.000	.296	(1.00)	.00	.00
D	(.955)	.000	.296	(1.00)	.00	.00
E	(.955)	.000	.296	(1.00)	.00	.00
F	(.955)	.000	.296	(1.00)	.00	.00
G	.402	(-.897)	-.183	.00	.00	(-1.00)
H	(.817)	.442	-.372	.00	(1.00)	.00
I	(.817)	.442	-.372	.00	(1.00)	.00
J	(.817)	.442	-.372	.00	(1.00)	.00
Est. total variance	61.4%	30.0%	8.6%	40.0%	30.0%	30.0%
Interpretation	Moderation	Extremism		5–scorers	4–scorers	1/7–scorers

(NB Dominant membership of subjects in factions denoted by oblique loadings in parentheses)

Each column in the factor matrix indicates the degree to which the N subjects (rows) conform upon a particular factor. The first column represents the main factor–that which accounts for the largest amount of systematic variation in the data—as extracted from the correlation matrix of Table 6.2; and each entry, or loading, in the column expresses the degree of relationship between this factor and a particular subject. The criterion for deciding whether a loading is high enough to denote a meaningful relationship between the subject and a given factor, while often quite tenuous (Comrey, 1973, p.226), is in the present application quite straightforward. On each row of the factor matrix the highest loading is isolated. Although its absolute value may be relatively low, it is at least evident that a subject has more in common with the factor to which his highest row loading relates than to any other. This assumption may be tested in particular situations by comparison of between factor raw scores, as in Experiment 25.

As usual, the relationship between an individual subject and a factor may be inverse; and subjects with a high positive loading upon a factor, while enjoying a high degree of similarity to each other, are highly dissimilar to subjects with high negative loadings upon it. Factor I of the unrotated test solution, containing a majority of high positive loadings, may be characterised as one of Moderation, giving weight to the subjects with similar scores of 4 and 5 on the seven-point scales. (It is nevertheless a factor expressive of slight bias to the higher or 'anti' scores on the scales.)

Factor II in the unrotated matrix can be characterised as an Extremism factor, comprising high loadings by the subjects with the most radical scores of 1 and 7 on the scales. Consequently, unlike the first factor, these subjects do not all share the same sign: subject G is inversely related to the other two, and on this evidence alone it would appear that the factor represents more than one subject grouping for resolution. In all, the test data contains four clear sub-groups that the unrotated two factor solution cannot separate: subject G, an 'isolate' with a positive relationship to no other subject, forms a notional sub-group in its own right.

It will be noted that the third unrotated factor in the solution, as compared with the first two, contains only minor loadings for all of the subjects. At the unrotated level of analysis it thus has no particular meaning; and, if the purpose of the analysis were simply to delineate the general interrelationships in the data (as accomplished above in terms of 'moderation' and 'extremism') such a factor would not be reported, and the solution would be meaningfully terminated at the unrotated factor II. But if the objective is an interpretation of the distinct clusters of relationship among the data, a sufficient number of unrotated factors must be extracted for the eventual separation of related clusters in the next stage of the analysis.

Isolation of subject factions

The separate 'factions' or between subject clusters in the data are isolated by performing a final rotation procedure upon the factor matrix, yielding a rotated matrix as in Table 6.3 (b). Since the individual subject factions are likely to be correlated to various degrees, both positively and negatively, an oblique solution is desirable (cf. Harman, 1967). Oblique rotation of the three factor solution given in Table 6.3 (a) emphasises firstly the cluster of subjects C to F in the original audience—a faction of individuals with scores of 5 each. A further faction (Table 6.3 (b), second column) comprises the subjects scoring 4s; while the third faction continues to emphasise the three extremist subjects. In less artificial data it would have been worthwhile to extract a fourth factor, onto which the fourth 'faction' (Subject G) would have been isolated from the extremist faction at this stage. From normal data, containing within subject as well as between subject variance, far more factors may be extracted and rotated in any case until the overall group is finally sectioned into N sub-groups with one subject in each. In practice, therefore, a cut-off point must be established at the initial factor extraction stage in order to prevent the solution from becoming meaningless; and, as previously, the 'eigenvalue one' criterion of Kaiser (1960) may be used for this purpose.

But it must be used with caution. For, if highly meaningful minority factions can fall beyond the level it prescribes, 'eigenvalue one' may in particular faction analytic situations prove too stringent. If this criterion had been uncritically applied in the present context, for example, the extraction of unrotated factors would have been prematurely terminated after the second. Though the criterion may be adopted for expedient reasons, it will always be wise to inspect the signs of high loadings on each rotated factor in order to determine whether the solution to that stage resolves as much of the interfaction conflict as is required, or whether further factions should be isolated. If individual subjects have high loadings on the same rotated factor and the same sign, the faction that they form within the total mass may be considered integral.

Seen in this light, the between subject correlation coefficient on which this analysis is based may be termed a *coefficient of cohesion.* A thoroughly cohesive mass in these terms is one shown to comprise but one main faction with no residual variance unaccounted. The percentage of total variance accounted for by the factions comprising other types of mass may be estimated, as indicated in Chapter 4, by summing the squared loadings on each faction, multiplying by 100, and dividing by the number of subjects in the total population. As factions are derived by oblique rotation these estimates are not necessarily exact; however, they provide an adequate guide to the relative importance of each subject grouping in the population as a whole (cf. Rummel, 1967, p.467).

Experiment 25: Comparison of factions

A series of audience attitude studies is now reported in illustration of the faction analysis principle. In the present experiment the technique is used to expose the differences between separate factions in a subject mass.

Procedure

Two sixty second selections were collated, one from each of two separate national TV news bulletins videotaped during the previous week. Three news items were included in each selection, covering contrasting themes: (a) a domestic current affairs item, (b) a dramatic item, and (c) a human interest item. The young lady newsreader, new to the position at that time, was presented in mid shot and her image was faded to black at the end of each item. One piece of film with news-reader voice-over (approximately seventeen seconds) was used in the central item of each selection. Each selection was then shown to ten male and ten female students (N = 40 in all), and their reactions to the newsreader were measured on 50 semantic differential scales. In aims and execution the procedure is directly comparable with that reported in a study of newsreader credibility by Markham (1968).

Results

A two way analysis of variance (programme selection x subject sex) and subsequent multiple comparison tests were performed upon the ratings on each scale. In programme selection 2, the newsreader was rated significantly more PERSUASIVE than in selection 1 ($P < 0.01$), and in general she was regarded as significantly more POPULAR (P 0.01) and DEPENDABLE ($P < 0.05$) by male subjects than by female subjects. Programme/subject sex interactions occurred on the RELIABLE and BELIEVING scales (P 0.05 in each). The scores relating to the two news selections were then pooled for all but the persuasiveness scale, and a series of t-tests was conducted confirming the above (male vs. female) results.

Coefficients of cohesion were next computed between the 40 subjects, and a preliminary faction analysis (FnA) was performed upon the total population by the method outlined in the last section. A major attitude faction was isolated comprising 21 of the subjects and accounting for approximately 19.23 per cent of the total between subject variance. The 40 subjects as a whole were divided into five factions, including one isolate faction and accounting for approximately 45.4 per cent of the variance overall: clearly a subject mass far less cohesive than, for experimental purposes, one would employ by choice.

In view of the analysis of variance results the male and female sub-groups (N = 20 in each) were next subjected to faction analyses in their own right. The male sub-group on this basis divided into three approximately equal attitude factions accounting overall for approximately 45.37 per cent of its total variance; while the female sub-group proved far more cohesive, dividing into two factions accounting for approximately 50.97 per cent of its variance (Table 6.4). The main female faction contained 16/20 subjects.

The raw attitude scores of the main male and female factions (N = 7 and 13 respectively) were next compared via a series of t-tests, one for each bipolar scale. The two scales observed significantly to differentiate male and female subjects before FnA (concerning perceived popularity and dependability) were found to do so again, and in the same direction. However, significant male/female differences were now observed on 9 additional scales as shown in Table 6.5. On all but one HUMOROUS/SERIOUS) the male viewers were more favourable in their ratings (of the female newsreader) than the female viewers.

A final series of t-tests between the scores of the three approximately equal male factions on each scale revealed 26 significant effects, classifiable as follows:

Table 6.4

Male and female factions in ratings of a female newsreader – Experiment 25

Subject	*Male factions* I	II	III	*Female factions* I	II
1	.176	(.761)	–.243	(.494)	.310
2	–.361	(.686)	.287	(4.77)	.422
3	(.506)	.222	.214	(.733)	.041
4	.355	.159	(.372)	–.222	(.876)
5	(.625)	–.183	.212	.309	(.524)
6	.283	.029	(.594)	(.700)	.088
7	.139	.205	(.634)	.122	(.658)
8	.112	(.640)	.212	(.767)	.009
9	.376	.219	(.420)	(.793)	–.001
10	(.652)	–.025	.093	(.710)	.077
11	.325	(.334)	.211	(.626)	.047
12	–.120	–.085	(.810)	(.421)	.419
13	(.683)	.113	–.165	(.873)	–.082
14	(.746)	–.019	–.045	(.463)	.332
15	.004	(.750)	–.022	(.823)	–.090
16	(.503)	.224	.333	(.834)	–.170
17	.186	.306	(.539)	.255	(.574)
18	.275	.164	(.471)	(.748)	.043
19	.194	(.473)	.210	(.381)	.376
20	(.641)	.089	.000	(.746)	–.025
Est. total variance	17.8%	13.8%	13.8%	38.1%	12.8%

(NB Subject representation in factions denoted by loadings in parentheses.)

Table 6.5
Differences between male/female ratings of a female newsreader before and after faction analyses – Experiment 25

Scale	Before faction analyses			After faction analyses		
	Mean scores			*Mean scores* (main factions)		
	males (N=20)	females (N=20)	t	males (N=7)	females (N=16)	t
FRIENDLY	2.65	3.35	–1.55	1.43	3.50	–4.30†
INTERESTING	2.85	2.85	0.00	1.57	2.62	–2.41*
POPULAR	2.75	3.70	–3.01†	2.86	3.81	–2.20*
FAIR	2.75	3.20	–1.25	2.00	3.37	–3.13†
DEPENDABLE	2.90	3.80	–2.68*	2.00	3.81	–4.27†
PRECISE	2.15	2.65	–1.17	1.29	2.31	–2.17*
HUMOROUS	5.40	5.00	0.91	6.29	4.75	2.82†
STABLE	2.45	3.00	–1.45	1.57	2.81	–2.55*
GENTLE	3.05	3.45	–0.99	2.71	3.75	–2.21*
SIMPLE	3.70	3.45	0.55	2.29	3.62	–2.60*
MEANINGFUL	3.15	3.35	–0.66	2.29	3.50	–3.13†
df			38			21

(NB Lower means denote more positive ratings; significance levels: *5%, †1%)

(a) Faction I rated the newsreader more favourably than Faction III on 10 scales, exclusively identified with professional qualities.

(b) Faction III rated the newsreader more favourably than Faction II on 6 scales, exclusively identified with personal qualities.

(c) Ditto (5 scales): Faction I more favourable than Faction II.

(d) Ditto (2 scales): Faction III more favourable than Faction I.

(e) Faction II rated the newsreader more favourably than Faction III on 3 scales, tentatively identified with professional qualities.

These analyses, combined with an inspection of the mean scores for each faction, suggest Faction I to have been predominantly motivated by a favourable regard for the newsreader's professional qualities and Faction III by one for her personal qualities. Faction II emerges as generally moderate: i.e. indifferent or ambivalent (see Chapter 4).

The general incohesiveness of the total subject population (N = 40) is thus largely due:

1 To sex differences between viewers' ratings.

2 To differences in the judgement criteria applied by individual male viewers.

Discussion

Comparisons of the major attitude factions within male and female audiences of a female newsreader (a) clarified sex differences observed between their populations as a whole, and (b) indicated the use of different judgement criteria by individuals.

Our immediate interest is less in the exact substance of these results than in the faculties they reveal in the faction analytic technique. In the case of (a) the role of FnA has been to amplify a previously apparent difference between subjects quite markedly; while in (b) it has been to reveal that judgement criteria whose precise origins were previously obscured can be due to the differing psychological outlooks of individual subjects. Thus, if preliminary analyses indicate a tendency towards varieties of social and/or psychological conflict within a population, and the number of subjects is large enough, FnA may be used to clarify the issue firstly by 'screening out' the subjects whose scores are relatively deviant, and then by 'homing in' on subjects whose scores are relatively uniform. Indeed the technique is as likely to emphasise the essential similarities between individual subjects and subject populations during the first of these two procedures as it is via the second; and in both the results' generalisable significance will naturally depend on the degree to which individual factions predominate within the population from which they originally derive. The significance of one faction's dominance over another may be assessed by a comparison of the percentages of total between subject variance for which they individually account (see Experiment 27).

Specific implications of the above male/female comparison concern possible relationships between the 11 scales significantly affected. On all but one of the scales (Table 6.5) the main male faction's scores are significantly nearer to the adjectival pole considered 'positive' in the original scale construction. On the evidence of an overwhelming male preference for the person perceived, therefore, we may conclude the positive/negative polarities of the HUMOROUS/SERIOUS scale to have been wrongly coded, and that, in the present newsreading context at least, seriousness has proved a greater virtue than levity. Though we have not emphasised viewers' reactions to perceived humorousness in TV communicators previously, the result may certainly be taken to support the strong previous evidence for audience attributions of 'benevolent insipidness' discussed in Chapter 5. (In presentations seeking to appear fully credible in professional terms, it is therefore quite likely that an attempt on the performer's part to appear humorous and personable will generate conflicting feelings towards him.) But, since it must be agreed that the positive/negative connotations of HUMOROUS/SERIOUS are at best relatively ambiguous, we would be loath to regard an inverse relationship between the scale and others as necessarily 'dissonant' in the manner this suggests. Just as ambivalence has been shown capable of confusion at the semantic differential midpoint with indifference (see Chapter 5), so evidence of attitude conflict may be wrongly inferred from inverse relationships due simply to the faulty coding of an ambiguous scale prior to analysis.

Future modifications of the semantic differential technique along the lines recommended by, for example, Kaplan (1972; see also above) may ease this problem by exposing the ambiguity of a scale in terms of specific response measures. Otherwise the significance of an inverse relationship between scales must be judged, by researchers and producers alike, pragmatically in the context of the particular programme attributes and objectives prevailing. Fortunately oversights in the directional coding of scales prior to faction analysis are of no immediate consequence, in view of the technique's dependence on a coefficient responding exclusively to between subject magnitude variance. Once the major sources of between subject variance have been rationalised in the manner illustrated above, coding criteria can be checked in the context of within subject factor analyses performed upon the data of the individual factions in the conventional manner. These analyses in turn yield the attitude patterns on which hypotheses concerning a faction's intrinsic characteristics depend.

Experiment 26: Analysis of faction characteristics

In order to clarify the qualities that the individual members of an attitude faction are each presumed to reflect, factor analyses may be conducted of the type used in Chapters 4 and 5 (R-technique). In the process, ambiguities as to the positive/negative polarity of a construct such as HUMOROUS/SERIOUS are indicated by the signs of the loadings on individual factors; errors in the construct's initial coding may then be corrected by a reflection of the signs.

Procedure

Using the data of Experimentj25, R-factor (within subject) analyses were performed upon the members of the main male faction (a), and upon the main female faction (b). For purposes of comparison with (a) and (b) respectively, further analyses of the same type were performed on the total male sub-population (c)–i.e. prior to faction analysis–and the total female sub-population (d). Finally, identical analyses were performed on the combined male and female population as a whole (e), and its main faction (f), for comparison with each other. The latter two analyses allow for a direct comparison of the data with those collected under similar circumstances by Markham (1968).

Each analysis involved a principal factoring technique as used previously. In view of the unusually large number of rating scales used–50–a series of higher order analyses was also conducted (cf. Rummel, 1970, Chapter 18) intended for publication at a later date: for this reason an oblique method of rotation was used (Harman, 1967) allowing the major attitude clusters in each factor solution to be highlighted despite marginal relationships between them. Since the subjects belonging to particular attitude factions are by definition likely to share highly similar scores rendering iteration impossible, a non-iterative method of factoring was selected. The extraction of unrotated factors was continued to the 'eigenvalue one' criterion; and the threshold for significance of factor loadings was set, as previously, at plus or minus 0.55.

Results

The analyses of ratings of the female newsreader by the total population (n = 40) and its main faction (N = 21) are compared first. The first three factors in each, both unrotated and oblique, are presented in Table 6.6. Before FnA, the most prominent pattern of ralationships in the data occurs on unrotated Factor I between attitude scales identified with perceived Mastery (i.e. expertise, strength, interest value and straightforwardness) and calmness (Poise?). Inverse relationships between these scales and others concerning perceived modesty and placidness brand the factor as moderately complex, and as reflecting professional attributions with a hint of dissonance. (Since it is difficult to imagine that attributes of immodesty and argumentativeness can ever be unambiguously positive, we do not suppose the above inverse relationships to be due on this occasion to faulty coding of the scores.)

Unrotated Factors II and III underlying the total population's attitudes each reflect more personal qualities (e.g. gentleness and pleasantness), though the psychological distinction between these factors is unclear. Interpretations of the corresponding oblique solution is equally awkward. The analysis of the same population's main faction, however, yields three primary factors which, at least after rotation, have a more distinct set of meanings altogether. These, indicating the use by faction members of judgement criteria relating to Integrity, Professionalism and Empathy (see Table 6.6), coincide exactly with the meanings due to the three major rotated factors reported by Markham (1968) despite the use in the present experiment of far fewer subjects and a partially different set of adjectival scales.

When the ratings of the male population (N = 20) are analysed, factor interpretations are once again problematic. Poise and Integrity factors are tentatively identified, though as above it is not until the population's main faction (N = 7) is analysed that a substantial number of high loadings on unrotated and oblique factors alike is observed, rendering the meanings of each factor apparent. Each of the primary factors observed at this stage, however, is highly complex and characterised by extreme inverse relationships between, for instance, perceived strength and impartiality, vigorousness and broadmindedness. Though the number of subjects in this faction clearly prevents us from drawing general conclusions from its characteristics at this stage, the pervasive inconsistency of its attitudes is notable in relation to effects observed in Experiment 28. Similar conflict is found, moreover,

Table 6.6

General population R-factors before and after faction analysis – Experiment 26

		Unrotated factors	Total variance	Oblique factors		Sum of squares
Before faction analysis (N = 40)	I	*Professionalism vs. (?):* PRECISE (.731), POSITIVE (.724). SHARP (.716), CALM (.635), EXPERT (.626), IMPORTANT (.622), MODEST (–.618), STRONG (.615), INTERESTING (.589), STABLE (.567), PLACID (–.562), STRAIGHTFORWARD (.553)	17.5%	*Integrity:* DIRECT PRECISE	 (.676) (.582)	2.684
	II	*Personal:* GENTLE (.653), FRIENDLY (.618), SINCERE (.615), TOLERANT (.588), REASONABLE (.563)	11.2%	*Personal:* HUMANE MODEST HONEST	 (.733) (.730) (.674)	3.394
	III	*Empathy:* PLEASANT (.626)	6.6%	*Empathy:* WARM PLEASANT LIKEABLE	 (.858) (.777) (.753)	3.338
After faction analysis (N = 21)	I	*Integrity:* TOLERANT (.867), RELIABLE (.719), FAIR (.706), IMPORTANT (.623), SINCERE (.614), MODEST (.598), HONEST (.590), STABLE (.552), STRONG (.552)	16.6%	*Integrity:* FAIR IMPORTANT RELIABLE REASONABLE	 (.838) (.790) (.568) (.557)	3.704
	II	*Integrity vs. (?):* EXPERT (.781), GENTLE (.625), CALM (.622), DEPENDABLE (–.617), CONFIDENT (.615), DIRECT (.585)	12.2%	*Professionalism:* CONFIDENT CALM EXPERT	 (.948) (.766) (.730)	3.954
	III	*Empathy:* LIKEABLE (.756), WARM (.733), PLEASANT (.731)	9.4%	*Empathy:* PLEASANT WARM LIKEABLE	 (.901) (.849) (.654)	3.209

(NB Only those loadings greater than ± 0.55 are tabulated)

in the ratings yielded by the other two male factions. (The first three oblique factors, underlying the main male faction's attitudes are presented in the context of Experiment 28: see Table 6.8.)

The corresponding factors revealed by analyses of the female subjects' ratings, on the other hand, are all completely consistent (see Table 6.8); and they may tentatively be identified with Empathy, Integrity and Mastery criteria similar to those underlying the judgements of the total population consensus of N = 21 (see Table 6.6). This interpretation is unchanged by Fna—not surprisingly in view of the marked dominance of the main female faction (N = 16) within the female population as a whole (N = 20) and *ergo* the latter's high degree of cohesiveness. Since the female population is manifestly so much more cohesive (thus more deadly . . ?) than the male, it is no more surprising that female subjects predominate within the main faction of the total population (N = 13/21), nor that the major consensus of attitudes in the total population is governed, in effect, by the females' coherent influence.

Discussion

Inspection of the major factors underlying the attitudes of main audience factions towards a female newsreader (a) revealed patterns not apparent from analyses of their populations as a whole, and (b) indicated sex differences in judgement consistency.

The experiment's primary objective—the identification of characteristics uniting the members of individual attitude factions—has been fulfilled by the observation of particular patterns in their ratings. Once again the FnA technique has served to clarify a variety of between scale relationships by highlighting alternative sources of variance in the data for independent inspection subsequently. Thus, after FnA of the ratings given by a (male and female) student population in response to a female newsreader, three previously indistinct judgement criteria have been identified; their correspondence to the criteria extracted from audience ratings by Markham (1968) suggests that the role of FnA in this case has been to produce a set of results of greater generalisability than could have been derived from the sample otherwise. By providing a criterion for screening out the between subject variance in the population, the technique has allowed the within subject effects characterising it to emerge relatively unconfounded.

The judgement criteria used by the female subjects in this situation were essentially the same as those consonant criteria underlying the main consensus of attitudes in the population as a whole. In contrast, those of the male population—once again clarified by FnA—were characterised by a high degree of conflict. In the course of this and the previous experiment, not only have three separate male factions been identified (between subject conflict) but also a high degree of inconsistency within the scores of each one (within subject conflict). No general conclusions are drawn as to the psychological reasons for these effects at this stage, though two specific predictions are indicated for further testing:

1 That newsreaders receive more uniform (i.e. cohesive) ratings

from viewers of the same sex as themselves than from viewers of the opposite sex.

2 That newsreaders receive more consistent ratings (on various scales) from individual viewers of the same sex as themselves than from individual viewers of the opposite sex.

We explore these predictions, and simultaneously conduct further tests of the FnA technique, in Experiments 27 and 28 respectively.

Experiment 27: Sex bias and group consensus

Having obtained evidence for attitude effects due to between subject and within subject variances independently, it is necessary for us to propose terminological distinctions by which to discuss them separately henceforward. When different attitude patterns were described hitherto (primarily in Chapter 5) their bases were unspecified; the terms used with respect to the extent of agreement vs. conflict characterising them (i.e. consonance and dissonance) therefore related generically to either one of them or to both at once. The specific terms used to describe the two types of effect henceforward are as follows:

Consensus—the extent of attitude agreement between subjects (each population faction represents one consensus): measured in terms of the total between subject variance for which the faction accounts. This usage is consistent with occasional uses of the term to this point and with its citation as one of the main tenets of attribution theory by Kelley (1967). High consensus contrasts with low consensus.

Consistency—the extent of attitude agreement within a subject's ratings between scales: measured in terms of the relationship between scales on the attitude factors of an individual faction. Throughout the present research this usage thus implies consistency of judgement criteria and not, as in other attribution contexts discussed by Kelley (1967), consistency over time or modality. The antonym is, of course, inconsistency.

The generic term for these phenomena retains its previous meaning; thus:

Consonance—the extent of attitude agreement due to between subject and/or within subject factors (i.e. not applied to specific within faction effects): measured factor analytically as in Chapter 5.

Other terms with a fixed meaning are:

Cohesiveness—the extent of predominance by one population faction among several: measured by comparisons of consensus.

Generalisability—the extent to which effects pertaining to one subject or faction may be abstracted for the purpose of interpreting others: estimated in relation to population cohesiveness.

The scope for measurement of these terms offers various possibilities for the further study of media treatment and audience variables. In the present experiment, for instance, we consider the relationship of TV viewing consensus to the audience factor of sex bias predicted by the results reported earlier in the chapter.

Procedure

A replication of Experiment 25 was conducted using 40 new student subjects, and a second set of national TV news items presented by a different newsreader; the latter was male and generally well known in contrast to the newsreader in the previous experiment. In all other details (scales used, analyses conducted, etc.) the two experiments were identical.

Results

Two way analyses of variance on the 50 adjectival scales yielded a significant between programme difference on the RATIONAL/INTUITIVE scale ($P < 0.05$), and subject-sex differences on eight others: as in Experiment 25, the newsreader was more favourably rated by the male viewers (i.e. significantly more HUMANE, PLEASANT, FRIENDLY, KIND, SHARP, WISE, STRONG and VIGOROUS: $P < 0.05$ in each). a result precluding the possibility that sex of viewers and newsreader interact with respect to simple preference. T-tests between male and female populations conducted before and after FnA indicate the exclusion of deviating subjects from the analyses to have rendered male and female viewers more similar in their attitudes to the male newsreader than was previously apparent: the main male and female factions were significantly differentiated by fewer scales than were their populations as a whole. The main male faction regarded him as more PRECISE ($P < 0.01$) than did the main female faction, and more PROFOUND, IMPARTIAL and UNEMOTIONAL ($P < 0.05$ in each).

The male and female populations were also very similar in this experiment with respect to group cohesiveness. Each of them was subdivided by FnA into three factions, including main ones accounting for 31.18 per cent and 30.70 per cent of their total variance respectively. The main male faction contained 14/20 subjects and the female faction 13/20. The consensus estimates for male and female viewers in the two experiments are presented in Table 6.7; and via the chi-square test (one tailed: Siegel, 1956, pp.107-9) a significant crossover interaction is observed between them ($\chi^2 = 3.424$, df = 1, $P < 0.05$). The main faction of the newsreader's audience as a whole comprised 19/40 subjects (9 male, 10 female) and accounted for an estimated 15.30 per cent of the total variance.

Discussion

Comparisons of the major attitude factions within male and female audiences of a male newsreader (a) accentuated similarities between their populations as a whole, though (b) provided evidence sustaining a marginal relationship between sex bias and group cohesiveness.

The results clearly fail to sustain outright the prediction generating this

Table 6.7

Consensus estimates (%): viewers x newsreaders – Experiment 27

Viewers	*Newsreaders*	
	Female (Experiment 25)	Male (Experiment 27)
Female	38.13	30.70
Male	17.83	31.18

experiment—that newsreaders receive more cohesive ratings from same-sex viewers than from other-sex viewers (see Experiment 26). The statistical significance of the crossover interaction between consensus estimates in this connection derives entirely from the difference between male and female consensus regarding the female newsreader in Experiment 25.

Nonetheless, the fact that the attitudes of female viewers towards a female newsreader should emerge as significantly more cohesive than those of male viewers merits attention. That the females have tended to react less favourably than the males towards her has already been established (see Table 6.5); and from Experiment 26 it is evident that they are also more consistent in their expression of attitudes towards her. It may readily be inferred that the female viewers' between subject consensus and their within subject consistency are due to a common psychological influence not shared by the male viewers: and it is further logical to presume that the effects of a sex bias of this type are at their strongest when, as in Experiments 25 and 26, the person perceived is little known and its qualities 'maximally ambiguous', cf. Scheff, 1973 (see also Experiments 12 and 20).

It is thus understandable that in reactions to the TV material featuring a well known newsreader the hypothetical sex bias upon group consensus is not observed: the less ambiguous nature of his performance has in theory obscured any effect of this type that might otherwise have been obtained. However, the lack of an effect of sex bias upon between subject variance does not imply that it may not influence the within subject patterns denoting attitude consistency (see second prediction, Experiment 26. The next experiment checks on this possibility.

Experiment 28: Sex bias and individual consistency

The specific prediction was as follows: that newsreaders receive more consistent ratings on various scales from same-sex individuals than from other-sex individuals. In order to test it we now replicate Experiment 26, using the ratings of the male newsreader collected in Experiment 27.

Procedure

R-factor analyses of these data were conducted as in Experiment 26. The solutions for the general population and the separate male and female sub-populations were examined before and after FnA as previously.

Results

In the solutions relating to newsreader ratings by the general subject population, the three factors denoting Integrity, Professionalism and Empathy judgements are clearly evident once again. As in Experiment 26 the interpretation of these factors is greatly facilitated by increases in the size of a high proportion of factor loadings after FnA. Also as in Experiment 26, the attitude patterns underlying ratings by the male and female sub-populations differ appreciably. Thus analysis of the main male faction (N = 14) yields a set of essentially consistent primary patterns emphasising separate assessments of 'inner' qualities, Professionalism and Empathy. Primary patterns underlying the main female faction's ratings, however, (N = 13) feature high inverse relationships involving perceived vigorousness and strength as in the other-sex faction in Experiment 26: these effects persist even after factor rotation. Table 6.8 summarises them and allows for their comparison with the corresponding results obtained in Experiment 26.

Discussion

Inspection of the major factors underlying the attitudes of main audience factions towards a male newsreader (a) revealed patterns not apparent from analyses of their populations as a whole, and (b) confirmed sex differences in individual consistency.

Though the newsreader in the present experiment was of the opposite sex to the one viewed in Experiment 26, it is evident that both have received more consistent ratings from individual viewers of the same sex as themselves than from those of the opposite sex. Precise quantifications of the attitude relationships revealing these effects will derive from the cue summation model propounded in Chapter 4: as indicated in Experiments 19 and 20, the model yields measures predicting the perceptual ambiguity of visual information and the systematic impact of different degrees of ambiguity on attitudinal and behavioural measures. It suggests that when ambiguity is maximal general conflict within a subject group's reactions will also be maximal: that conflict is determined by ambiguity and varies according to the level at which it is perceived. But, in the present experiment, judgement conflict—or more specifically the (within subject) consistency variety—is also influenced by sex bias; and we may deduce that in individual subjects the latter may affect perceived ambiguity and the consistency of judgements alike. A specific inference to emerge from this experiment is thus that TV newsreaders appear more ambiguous to viewers of the opposite sex than to those of the same sex—

		Female newsreader (Experiment 26)	Sum of squares	Male newsreader (Experiment 28)	Sum of squares
Main male factions	I	*General (D):* PRECISE (.980), S'FORWARD (.951), CALM (.931), STABLE (.931), INTERESTING (.829), PROFOUND (–.776), SUPERIOR (–.774), LIKEABLE (.697), HUMOROUS (–.694), BELIEVING (–.629), POSITIVE (.594), PLEASANT (.575)	9.527	*Inner:* SINCERE (.953) REASONABLE (.824) PROFOUND (.673) GOOD (.672)	4.632
	II	*Mastery vs. Integrity:* STRONG (–.933), IMPARTIAL (.922), MEANINGFUL (.889), SPONTANEOUS (.881), TOLERANT (–.828), EXPERT (–.770), DIRECT (.719), INFORMED (.659), POSITIVE (.590), PERSUASIVE (–.563), DEPENDABLE (.557)	9.284	*Professionalism vs.(?):* CALM (.836) CONFIDENT (.732) PERSUASIVE (–.688) RELAXED (.613) EXPERT (.561)	4.615
	III	*General (D):* VIGOROUS (.909), BROADMINDED (–.903), WISE (–.845), IMPORTANT (.832), FAIR (–.753), UNEMOTIONAL (–.719), PLEASANT (.646), REASONABLE (–.623), RATIONAL (.616), SUBTLE (.580), HUMOROUS (–.573), GOOD (.568)	8.346	*Empathy:* HUMANE (.812) TOLERANT (.797) POPULAR (.692)	3.966
Main female factions	I	*Empathy:* POPULAR (.757), REASONABLE (.704), FAIR (.669), IMPORTANT (.652), BROADMINDED (.630)	4.064	*Personal* WARM (.912) PLEASANT (.858) LIKEABLE (.845) GENTLE (.790) KIND (.770) BELIEVING (.767) SINCERE (.753)	6.738
	II	*Integrity:* DIRECT (.866), PLACID (.803), SINCERE (.634)	3.977	*Mastery vs.(?):* VIGOROUS (–.803) PROFOUND (–.688) CALM (.595) RATIONAL (.567) SHARP (.566)	4.595
	III	*Mastery:* INTERESTING (.929), WARM (.630), GENTLE (.618)	3.533	*Mastery vs.(?):* STRONG (.815) POPULAR (.754) INTERESTING (.680) PLACID (–.614) GOOD (.612)	4.083

(NB Only those loadings greater than ± 0.55 are tabulated; (D) denotes dissonant factor) a conclusion for which the use of faction analysis has been primarily responsible. Further experiments would help to determine whether similar effects occur in wider communication contexts.

This effect is clearly only one of countless psychological phenomena which may be identified with the mediation process via the use of appropriately sensitive analytical techniques. We report it purely in order to illustrate the uses of faction analysis in studies attempting to predict the complex interactions between communication treatments and audience differences. In concluding this chapter we discuss FnA's functions in more detail: firstly, however, we apply it once more in checking a final prediction regarding within and between subject effects in the present data.

Experiment 29: Sex bias vs. familiarity

It is apparent that the consistency of an individual viewer's attitudes to a TV performer and the consensus of opinion between different individuals in the same situation can be differently determined. In Experiments 26 and 28 consistency effects have been attributed to viewing sex bias. They derive, it appears, from ambiguities in the performers' image as perceived by individuals, and they remain constant despite differences between performers' familiarity to their audience generally. In Experiment 27, on the other hand, a consensus difference was relatively slight when performer familiarity was high; and the general potential of the sex bias factor in relation to group consensus is therefore as yet unclear.

In the present experiment we check whether, by controlling the performers' familiarity to their subjects, we may produce sex bias effects upon attitude consensus and consistency alike. Only if this should prove to be so may we predict the susceptibility of attitudes to both types of effect by sex bias in different viewing situations henceforward. The specific prediction for immediate testing is that when a male and a female newsreader are both unknown to their viewers the interactive effect between group consensus estimates by which the (between subject) sex bias is established will be greater than that observed when one of the newsreaders is well known as in Experiment 27. A positive result on this occasion would suggest that the effects of sex bias on consensus, while predictable in certain situations, yield easily to the effects of familiarity whereas consistency effects—possibly through greater robustness—do not.

Procedure

Ratings of an unknown male performer were obtained from 22 male and 22 female subjects for comparison with those of the little known female in Experiment 25. The experimental design, scales and administration were as in Experiments 25 and 27.

The videotape material had been used previously as the unedited condition of Experiment 5, being of a two camera (three and a half minute) interview about poetry; the performer rated was the interviewer. The choice of this material in the present context was governed, firstly, by the performer's total unfamiliarity to his audience and, secondly, by the relative dearth of mediating effects upon his earlier impact. The latter, associated independently with a high degree of performance 'motive' (see Experiment 12) is assumed to mitigate against the likelihood that an interactive effect between consensus estimates will be observed as predicted. Though the material is not identical to that of Experiment 25 as regards performance style and subject matter, its qualities in the above two respects render it adequately comparable with Experiment 25 for the present purpose.

Results

Faction analyses of the new male and female audiences yielded main consensus estimates of 29.16 per cent and 18.61 per cent respectively. When compared with the corresponding estimates yielded by Experiment 25, these indicate a significant interaction between subject-sex and performer-sex as predicted ($\chi^2 = 7.717$, df = 1, $P < 0.01$). The reactions of the same sex subjects were more cohesive than those of the other-sex subjects in both experiments; and the interaction itself is more pronounced than that observed in Experiment 27, consequently upholding the specific prediction made at this experiment's outset.

R-factor analyses of the main factions' attitude structures indicate a predominance of Integrity assessments with some inconsistency evident in both male and female subjects. As in Experiments 26 and 28, however, the other-sex subjects (in this case female: N = 8/22) demonstrate a characteristic orientation towards qualities of strength and vigorousness, both of which conflict with personal qualities. The same-sex subjects (N = 14/22) do not share this bias.

Discussion

Comparisons of the major attitude factions within male and female audiences of a male interviewer confirmed predictive relationships between sex bias, consensus and consistency in conditions of unfamiliarity.

Specific effects of sex bias upon attitude consistency are observed for the third time in succession, and their potential importance in TV news and discussion contexts is clearly indicated. Having this time controlled the familiarity variable, we have also been able to predict a significant effect by sex bias on group consensus; and it now remains to be seen whether such effects also operate in general communication situations.

Certainly, when a person in any context is unfamiliar to those perceiving him, marginal psychological biases have a prime opportunity to affect attitudes towards him (cf. Festinger, 1954). In the present chapter this has evidently been the case at both between and within subject levels; indeed the effects of bias upon within subject consistency have appeared highly predictable whether performer familiarity is minimised or not. The effects on between subject consensus, on the other hand, have proved less predictable owing it appears to the priority of familiarity effects over them. Greater familiarity with a person thus causes consensus and consistency to develop independently. If a person fulfils a recognisable stereotype or 'public image' already formed, individuals may be united in a nonetheless inconsistent impression of him: viz. in assessing the male newsreader in Experiments 27 and 28, female subjects were no less cohesive than the males though they were certainly less consistent.

How may these particular conclusions be applied in media practice? For example, how may TV producers avoid the sex bias effect in news broadcasting? All such results generate two alternative courses of action. Attempts may be made to control the effect either by modifying production stylistics–by, for instance, presenting male and female newsreaders side by side on as coequal a footing as possible; or by reducing the prejudices of the viewers–educating them in effect with regard to the basically irrational biases they harbour. In fact we would be reluctant generally to recommend the former course, pandering as it does to psychological differences with no guarantee of overcoming them: in particular situations it may also be wholly impracticable. A successful effort to help viewers recognise the groundless nature of many of their perceptual and judgemental prejudices, on the other hand, would speedily render the need for awkward changes of presentation stylistics redundant. Whatever the bias–whether originating in the physical image or the mental imagery–attempts to reduce it in practice may prefer to concentrate less upon the treatments to which it may be attributed than upon the judgement processes via which it takes hold (see Chapter 7).

A range of particular questions arises from these results. Does a performer's popular image develop primarily on sex based grounds, for example? Are same-sex subjects, in being more harmonious in their judgements, also more objective or less so? When in due course such questions may come to be investigated, it is evident that presentation factors, audience differences and basic stimulus familiarity may each exercise interacting effects to be borne in mind. Such interactions are not only in evidence throughout the present research but are also predicted by the three way scheme of Kelley (1967, p.195) whereby the entities perceived, the persons perceiving them and associated temporal factors each interact in the attribution process generally. Applications of attribution theory in a wide range of situations henceforward will undoubtedly help to clarify the n-dimensional origins of communication bias, both verbal and non-verbal, and they should be designed to take stock of its behavioural as well as attitudinal consequences.

The major object of all research intending practical benefits should inevitably be to increase predictability, initially in specific contexts and ultimately in general. In both respects, faction analysis may be claimed to have a useful future role. Owing to the technique's usage in this chapter, for instance, a particular tendency–unsuspected beforehand and thus objectively determined in the final analysis–has been recognised towards conflicting uses of the Mastery criterion within other-sex judgements (i.e. concerning performers' strength and, by association, vigorousness). This evidence takes us beyond the level at which subjects' responses are differentiated on the independent basis of, for instance, sex alone, and may generate new insights into the perceptual and judgemental tendencies with which differences in sex are generally associated. These insights can in turn help in developing a more fundamental and pliant set of criteria by which to differentiate our experimental subjects during investigations of the massed effects of communication media.

Factions and generalisability

The particular aspect of faction analysis rendering it useful in audience prediction during this chapter has been its capacity to identify each main group of 'faction' or subjects with overall attitude profiles essentially similar to each other, though measurably different–and possibly in opposition–to those of the remaining subjects. The exclusion of deviating subjects from the experimental population 'screens out' the variance for which they are individually responsible, and which may otherwise diminish meaningful tendencies within the data as a whole. The technique's essential function is thus to provide criteria for improving that which Cronbach et al. (1963) and Kaiser and Caffrey (1965) have defined as psychometric generalisability. Comparing the members of a subject mass as to their uniformity on a given data base, it indicates (a) the extent to which the mass is consolidated in its attitudes or internally divided, and (b) the sub-group based effects with a generalisable significance in relation to particular treatments that would otherwise have gone unnoticed. When subjects are found via the technique to be highly cohesive, FnA's usage serves to confirm the criteria by which they were selected; when they are not, the different variables predicting membership of each faction may be determined *a postereori* via, for example, multiple regression analysis (Nie, Hull et al., 1975, Chapter 20). The criterion for separating a mass into factions in this way is purely statistical and requires no prior knowledge of the sources of variance influencing subjects in the task at hand; the whole procedure is accomplished, therefore, in a manner free from contaminating assumptions.

Based on a factor analytic technique with oblique rotation, FnA does not contain implicit measures of statistical significance or criteria denoting the reliability of the conclusions it indicates. Future incorporation of procedures for computing the exact variances accounted for by oblique factors–as by Hofmann (1975)–may certainly extend the technique's versatility in these respects. In situations where the effects of a previously defined variable are examined–of a particular style of communication or environment, for instance–a degree of subjectivity in FnA's usage may clearly be unavoidable. Effects upon group cohesiveness must be weighed against criteria which, if not independently available, must be defined intuitively. Furthermore, inspection of an audience's factions in the attempt to establish the most generalisable effects upon them can be frustrated when separate factions are similar in size. In this situation a process of successive replacement of the subjects displaying least definite faction loyalties may determine whether one faction should predominate; yet ultimately the decision as to which set of effects is generally the most meaningful may in part at least remain subjective.

A TV producer, for instance, wishing to establish the main effects of his programme on the audience overall, may discover two distinct factions underlying its responses, one accounting for 17 per cent of the between subject variance and the other for 13 per cent. Whether the effects to be observed on the former are generally more predictive than those upon the latter is clearly debatable; however, the question more properly to be asked in this situation concerns the reasons for their difference, and involves the comparison of factions via further methods. It is the 'hung-jury' or 'split-vote' situation that analysts of a plebiscite might wish to explain. To production staff involved in an election campaign, therefore, the objective definition of separate attitude factions within the audience may prove of great value; while the knowledge that as little as 13 per cent of the between subject variance underlying viewers' responses is of a particular type could prove highly encouraging in the context of one production, though deflating in another. Criteria for the meaning and generalisability of a faction's responses thus vary intuitively between different applied contexts.

In general contexts involving unpredictability and conflict, however, developments of the FnA technique and of the Q-factoring proposition it embodies may serve a wide range of purposes. In mass communication studies the application of Q-factoring, long established in experimental psychology, is perhaps overdue. Communication research has not traditionally concerned itself with group conflict nor with the inconsistencies of subjects' responses; and these must often have been overlooked by the empirical approaches adopted. Williams (1973) suggests that the linkage of communication science with the 'mass' concept has actually stunted its development as a useful and predictive force. Faction analysis, a technique for penetrating the mass in its most opaque and amorphous of forms, may help to disentangle causes and effects within it which were previously undisclosed.

The intention behind the technique is not that it should necessarily act as an end in itself, but that it should help to divide the empirical process into several unaccustomed stages. This is exemplified in the correlation coefficient on which FnA has been based—one which makes an unusual provision for inspecting between subject variance prior to the usual within subject calculations. Whenever comparisons of scores on a standard scale are to be made for similarity or, for example, test-retest reliability, the coefficient of cohesion should have uses in its own right. When the raw scores have been collected on incompatible scales or are standardised, however, the coefficient is clearly inappropriate. If it is logical for between subject and within subject variances to be considered simultaneously, a pattern-magnitude coefficient such as the intraclass measure of Kendall and Stuart (1961) would be more suitable. (See also Rummel's discussion of alternative correlation techniques, 1970, Chapter 12).

The need, prior to analysis, for an examination of the highly specific assemptions underlying many social science statistics is in fact all too rarely realised. Certain techniques gain in precedence over others not by virtue of their greater versatility but owing to the fame that they gain in popular applications. When their limitations in general contexts are eventually recognised they appear in contrast, but wrongly, to be outdated and blunt. Yet statistical techniques, like any tool, are blunted primarily by uncritical usage, against infrangible surfaces and using the bludgeon instead of common sense. Faction analysis, as a new device, embodies a set of procedures which in certain situations have an evident precision. In its own interests, however, the technique should be reappraised and modified carefully each time a situation calls out for it, so that it may retain its precision yet gain in flexibility.

7 Non-verbal mediation: a summary

This has been a report of the initial series of experiments conducted by psychologists at Liverpool and Lancaster Universities during a five year research programme. Based on a theoretical viewpoint discussed by Baggaley and Duck (1976) and further assisted by funds from the Social Science Research Council between 1976 and 1978, the work has examined some of the technical and perceptual factors influencing our reactions to TV and to the information we receive from it. Via tests of audience reaction to a wide range of presentations, the effects of TV have been ascribed to a complex interaction between many levels of information—both verbal and non-verbal—which at their most subtle are often unpredicted by production staff and audience alike. A prime aim of the work has therefore been to identify the distinct strands of information involved in this interaction with a sufficient precision to enable producers and viewers to take better stock of TV presentation effects henceforward.

In specific situations the credibility of persons seen on TV, and by implication of the views and information they convey, are found to be affected by unconscious biasing factors that once identified may certainly be controlled. Naturally the scope for abuse of the medium on this basis is as great as the opportunity to extend it beneficially. In this final chapter we therefore distil a series of guidelines from the work, each intended to suggest ways in which the persuasive effects of TV can be harnessed or, where necessary, avoided. Our major goal is thus a practical one; we hope that the producer of TV news or educational material will find it useful to refer to particular aspects of the work and to check his own understanding of presentation effects against those we have observed through experiment. If our results should indicate that a particular technique—concerning, for example, autocue usage or a certain style of performance—has a more subtle capability than he had originally suspected (and if this causes him to speculate about its effects in other situations) then we will consider the work to have been worthwhile. If, in addition, our general findings can assist TV viewers to learn or interpret information more effectively from it, we shall feel justified in continuing to study the effects of TV imagery in further contexts.

Untying the Gordian knot

Though we have striven to emphasise the practical implications of this investigation wherever possible, we have also been obliged to stress the numerous methodological decisions which have gradually led us to our conclusions. Interpreting our empirical findings within the framework of attribution theory (Kelley, 1967), we have been essentially engaged in charting the types of effect that may operate, independently or in complex interactions, when people are viewed on TV in different contexts; and we have thereby been involved in the dissection of a process, the untangling of a knot of variables rather than the inspection of any one variable in fine detail. The approach we have adopted has sought to uncover a little about a wide range of factors, and to keep the whole in view throughout, rather than a lot about a narrow section of the field in impractical isolation; and when specific sources of media effect are examined in other contexts it may be of benefit for their interactions with the range of variables detailed above to be considered in parallel.

An attribution theory approach to the hypothesised effects of televised violence, for instance, would ensure that these were viewed very much in relation to the independent factors with which they logically interrelate. The approach would presuppose that TV, as any agent capable of mediating ideas to the impressionable, has the potential to increase social violence or indeed to reduce it. Whether it is more or less capable of such effects than other agents would be demonstrated in the joint terms of:

1 Reactions to the wide range of alternative factors thought capable of such effects.
2 Reactions of different people to televised material.
3 Reactions by individual viewers to TV material over time.
4 Reactions to the different ways in which televised violence may be mediated.

Compare these questions with the analysis by Kelley (1967, p.194) of factors determining the aesthetic effects of a film. Evidence in each of these respects can prove useful in specific contexts, but, if the variation represented by all of them cannot be taken into account simultaneously by investigators, any attempt to draw fully generalised conclusions on the subject is premature.

The particular process studied in the current connection has been that of mediation. As an independent research focus, mediation and media studies refer to all those factors capable of influencing the passage of meaning from one point to another in either space or time. They thus involve a broader set of questions than the analysis of mediating technique alone can resolve; for mediation subsumes the reception as well as the despatch of information, and its study attempts to identify the systematic ways in which a message may be conditioned by both. It is also a more elaborate problem than necessarily presupposed in the study of communication processes at large. The student of linguistics, for example, or of the communication made possible through music or fine art, is generally satisfied to classify the meanings recognised at the despatch stage of a message only. The rules of musical harmony, as valid a subject for study by communication specialists as any, refer to artistic criteria formulated by the composer and critic, while leaving the psychological effects on those who hear it as a rather less tangible question for aestheticians to resolve. The study of mediation, on the other hand, attempts to embrace both sides of such matters; by emphasising the process generated *between* separate points or persons it forces us to define communication phenomena less in terms of the absolute qualities evident at either locus than of observable interactions between both of them. It marries the intentions behind a message with the interpretations of it and—depending on the investigation's applied purpose—it may view each in the light of the other at will.

The study of non-verbal mediation accordingly takes as its immediate problem the process relating the design and transmission of, for example, visual signals to the reactions of those who perceive them. Numerous similar processes may operate within a single communication context simultaneously; and the distinction between the concepts of mediation and communication in general become particularly useful the more that this is so i.e. when the messages conveyed are complex and their audience comprises many diverse elements. In seeking to study the influence of communication via TV we are faced with a complex problem par excellence; and we must be prepared to account for the separate mediating factors that interrelate within the process and for the interaction of external influences with them. The whole is a Gordian knot whose untying requires a set of correspondingly intricate research tools.

The qualities demanded of a research technique do not always become apparent until it is in use. If we were to begin the present research again, for example, we would certainly bear in mind principles that have emerged as the complexity of its subject matter has unfolded. Thus we would modify the device used for attitude measurement so that the ambivalence element within our data might be gauged more precisely (see Chapter 4): via an experimental tool such as that proposed by Kaplan (1972) we would retain the ability to measure the differential effects of TV presentation upon viewers' attitudes as above, but would add to this a more detailed understanding of attitude conflict. A further refinement to our technique would be to preface each analysis with an inspection of the factions comprising each subject population. The application of faction analysis to our data before their within group and between group characteristics are inspected by more normal methods plainly helps in unravelling the individual differences bearing on media effects (see Chapter 6). In this connection we would certainly be obliged to double the number of subjects typically investigated in each of our conditions: and, if we had applied this technique earlier than the present Experiment 25, it is unlikely that the available supply of subjects would have been sufficient for the relatively large number of experiments that, as it is, we have been able to conduct. On another occasion we must attempt to overcome this the most common logistical problem that social science researchers face.

With these qualifications we believe that the measurement tools applied during the present study may prove useful in communication and media research more generally. The customary focus of attention by researchers upon massed reactions to the communication media has generated a tradition which, in Williams' opinion (1973), casts little light on the 'specific modern communications conventions and forms' or on their effects. Anxious to define the general effects of media content and technology, analysts have overlooked the audience members' basic capacities to disagree, to respond to different aspects of the stimulus in different ways, and to form the ever shifting sub-groups that define a society's dynamism. They have emphasised the broad similarities between subjects at the risk of permanently obscuring the ambivalences of human reaction and the ambiguities of meaning with which it must cope. The theory of attribution can provide particular insights in this connection, predicting the supportive evidence that an individual may seek when ambiguous information is presented to him; and when audience judgements refer to the human source of this information it appears that four judgement criteria are generally applied, concerning separate levels of the performer's personal image as well as of his professionalism. Via factor analysis we may see the ways in which these separate criteria reinforce one another or conflict; and by identifying the presentation cues that affect their usage we may predict the summative effects of attitudes upon behaviour. The knot we must dwell on thus entwines the variables of attitude complexity, individual difference and mediating technique. Having distinguished these, we may begin to attribute rules of cause and effect to the mediation process just as attribution theory asserts that we interpret the world around us normally: 'This understanding is gained by way of a causal analysis that is "in a way analogous to experimental methods" . . . and that has the purpose of disentangling the independent factors present' (Kelley, 1967, p.194).

The notion that in coming to grips with his environment and with mediated versions of it Man learns to act as an experimentalist, might at one level suggest that he has much to gain from a study of the logical rules of scientific enquiry. Certainly, we may all profit from the ability to behave methodically at times of crises; but if we emulate the ways of scientists too closely we may find that for normal purposes the practice is restrictive. In order to cope successfully with experience we do not necessarily have to recognise all of the factors present within it at any one time; nor do we need to consider all of the possible interactions between separate factors before learning to control them slightly better than before. If our experience has taught us that individual elements in the environment have a certain potential for effects, we can often anticipate them in practice quite satisfactorily even before they become manifest.

Indeed the scientist—particularly if he has an essentially complex problem on his hands—may learn much from a study of the pragmatic stances adopted by Man the survivor in everyday life. From the phenomena he observes in one experiment and the patterns that emerge across several he can frequently gain a working knowledge of the problem he faces, even though the pedantic side of his nature urges caution. If his subsequent treatment of the problem then proves successful his intuition has all the vindication it requires. The skills and practice of medicine are founded on much imperfect wisdom of precisely this type; and the applied social sciences can provide society with far more substantial benefits via practical recommendations formed on an interim basis than they may ever yield by the slavish adherence to criteria for ultimate proof.

Many of the recommendations derived from the present research stem from a consensus of evidence reinforced by intuition for just this reason. However often a given effect may be observed and however clear its origins may be, situations will in any case always be encountered in which, contrary to prediction, the effect does not appear. When they are, the ability to swallow more idealistic criteria and to fall back on intuition may prove most useful; for it is clearly unadvantageous to delay sensible applications of a piece of research for the sake of minor exceptions to an academic rule. The critic of experimental methodology in the social sciences points to the inability of experiments to cater for all eventualities in arguing that their relevance in applied research contexts is strictly limited; and to a major extent the argument is correct. For even when a conclusion has been experimentally proved beyond all shadow of a doubt only the practitioner can determine whether or not it has a useful application. On this basis experimentalists who overstate the predictive merits of their methods are more than likely to receive the short shrift from practitioners that they deserve. If all that they attempt, however, is to warn the latter of possibilities, to pinpoint potential influences within the complex of factors they have considered so that fresh techniques may be developed to adapt or guard against them, the only criticism they need truly fear is that the practical lessons were already obvious.

To the researcher of media we therefore offer the following recommendations:

1. That experimental methods should unhesitatingly be applied in separating the threads of a complex phenomenon for theoretical purposes, but that the ultimate applied value of the research should be decided pragmatically.
2. That the initial hypotheses on which the research is based should be as adventurous as possible and its outcome at least partially unpredictable: since the balancing of all independent variables in a single experimental design is all but impossible, experimental resources are less profitably used in straight replications than in the investigation of at least one variable not controlled previously.
3. That the variables themselves should therefore be considered capable of alternative effects depending on the contexts in which they are manipulated.
4. That the empirical measures used should relate as specifically as possible to the problems and objectives of the producer and/or audience; and that they should be capable of modification as more fixed criteria in these respects become apparent.
5. That the framework for interpretation of the results should be sufficiently fluid to permit retrospective assessments quite counter to any that may have been anticipated: i.e. that interpretative as well as design criteria should be capable of perpetual modification as the research develops.
6. That in general a common sense balance should be adopted between the rigidity for which experimental methods are frequently criticised and the looseness identified with armchair speculation; in an ongoing series of studies problematic design criteria may be relaxed to a greater extent than otherwise, since the loose ends occasionally evident in individual experiments will be amply redressed by any effects observed consistently across the series as a whole.

In fact it is only via an ongoing series of studies liberally designed that the independent strands of complex problems (including those initially unforeseen) may be identified and also re-entwined for inspection once again in their natural interactive state. Within the context of a single methodology used continuously, many between study comparisons may be made that would otherwise be unacceptable. If, despite the use of a fluid empirical approach such as we have advocated, consistent trends indeed emerge across a series of studies, they may be concluded reliable. It then remains to apply them to a useful end.

Guidelines in TV communication

The applied purpose for which the present research has been intended is twofold: firstly, to assist producers of TV to recognise and anticipate its effects when using it to educate and inform; and, secondly, to assist viewers in identifying the effects to which they may be subjected, whether beneficial or adverse. Some of the effects we have identified during the present research have naturally been more fickle than others; and the principles we derive below relate primarily to the general evidence observed across the series as a whole. They are intended to supplement the existing repertoire of guidelines borne in mind by TV users when designing or assessing televised material in specific situations.

For the producers of TV imagery we make five general points as follows:

1 Viewers clearly form very detailed assessments of the credibility of information presented on TV on the basis of extremely subtle visual cues. In assessing the extent to which audience responses may be influenced by particular production techniques, TV personnel should recognise the high priority of perceived credibility in the viewers' regard, and should continually scan the images they produce for cues capable of denoting it. Even the most tenuous of detail may cue specific attributions in ambiguous situations.

2 Highly consistent effects are generated by the performance style of the presenter and the apparent reactions of other persons to him. Both factors are predicted by attribution theory, and each of them may interact with other factors to produce a cumulative effect detrimental to the presentation's impact. Producers should pay close attention to this possibility in the direction of performers and in their usage of, for instance, visual cutaway shots.

3 The general imagery of TV is used by viewers in support of their assessments of a performer's Integrity, his Mastery over the information he presents, his personal warmth or Empathy, and his relaxedness or Poise. Attitudes formed according to any one of these criteria may complement or indeed conflict with those formed on other bases; and it should not be supposed that a competent performance in one respect is regarded by viewers as a criterion for a favourable assessment in general terms. Producers and performers should therefore concentrate on projecting the attribute(s) they consider most essential to the function and task at hand, rather than simply conforming to general stereotypes.

4 When the performer's prime function is to entertain, it is likely that viewers will attach a high priority to his personal Empathy and Poise. However, when it is to inform, viewers exhibit a consistent tendency to mistrust a friendly or polished performance and to downgrade the performer's perceived Integrity and/or Mastery as an apparent reaction against it. Similarly, a tense or even stern performance can earn their high respect. The outward attributes of a TV performer are evidently regarded as capable of misrepresenting his inner qualities; and attempts to personalise the reading of TV news, for instance, may well detract from the image of authority and expertise that it would otherwise deserve.

5 The tendency towards a conflicting audience reaction to the persons seen on TV is particularly marked in many situations where a reportive format is used and the camera is addressed directly. Viewers plainly recognise the essentially contrived nature of this type of behaviour, and are not inclined to believe that the performer is genuinely 'speaking to them'. The more a performer attempts to foster this impression the more he is likely to be regarded as an actor, and thus capable of feint. When simultaneously he uses an autocue or teleprompting device, his behaviour may be unnaturally constrained. Whenever the performance can be presented in an interview or discussion format, therefore, its impact is likely to be enhanced; for greater opportunity will be available for the audience to structure their independent assessments of it according to normal social criteria.

For suggestions regarding further practical options that he may take, the producer is recommended to consult the individual experimental sections above and the appraisal of evidence in Chapter 5. Thus the specific effects of visual background detail may be questioned, as can the importance of separate visual and aural image cues and the manipulation of cues in general by the selective use of camera and editing techniques. In putting forward these guidelines we at all times recognise that, in practice, TV direction involves spot decisions made spontaneously in situations that are often totally unpredictable. But the spontaneous act may nonetheless comprise carefully acquired skills of response, and its efficiency is perpetually increased as new skills are assimilated into the repertoire. The principles we have summarised here may assist in this process, informing decisions made by the producers of, for instance, educational TV and thereby increasing the options available to them in particular situations. In this way the instructional media generally may gather together a body of effective technical rules where few yet exist.

Attempts to control presentation impact raise obvious ethical questions. Desirable though it may be for educational practitioners to acquire new methods in this connection, it is essential that the viewer also should be made aware of the subtle effects of visual imagery upon him so that he may consciously resist them as he pleases. The need for public education in the skills of 'visual literacy' has been discussed by Baggaley and Duck (1976); via broadcast information concerning 'behind the scenes' technique and greater attention to media and communication studies in secondary and higher education, the intentions underlying media control may be clarified and unscrupulous ones recognised before they can take wing. The argument for early schooling on these issues gains some support from the observations in Chapters 2 and 5 above regarding the intense conflicts produced by TV imagery within the attitudes of children. The patterns underlying children's reactions to the mediation process may clearly be more elaborate than we generally suppose, becoming simpler as the child learns to recognise them, and in a non-ambiguous environment ultimately fading altogether as he bases his attributions more advisedly. Any attempts that can be made to assist the child's understanding of media effects will certainly be worthwhile. Doubts as to the ethical responsibility of research in the field may be allayed on the same basis.

Each of the conclusions drawn to the attention of TV production staff by our research may thus be rephrased for the viewers' benefit. While the producer may be urged to bear the psychological tendencies of his audience in mind when designing the images he presents to it, so also may the audience be encouraged to recognise that attributions based on mediating technique need by no means be reliable. TV viewers are certainly well able to recognise the sleights and contrivances of production technique in large measure for themselves–witness the sophistication of their response to reportive performance styles. But there is always room for them to assimilate new skills into their repertoire just as there is for the producer; and once both have acquired related skills, immense scope is indicated for new, more advanced communication methods in the media of the future: '. . . as long as the intentions of the media producers remain evident to their audience, and the latter's fluctuating needs are permanently clear to the broadcasters, the medium will convey a clear message, free from dynamic "noise", effectively transparent ' (Baggaley and Duck, 1976, p.172). The extra impetus to media development that may be gained from a greater technical understanding between producers and audience may, in future, surprise us. For in theory no limit can be placed on the extent to which images, whether aural or visual, may carry coded meaning. If their usage can be learned by sender and receiver alike even highly rarified codes can serve a valuable communicative function, as the final experiment suggests.

Experiment 30: The zoom effect

An attempt was made to test a speculation by Baggaley and Duck (1976, p.142) regarding the possible symbolic uses of the magnifying zoom lens. In 'zooming in', it was suggested, the ETV producer not only has a tool for the magnification of objects, but also one for directing attention to a significant point; while in 'zooming out' he can imply that the importance of points being made is, for the moment, diminished. The capacity of the zoom effect to carry various types of coded information, even unintentionally, has been indicated by Salomon (1972), and its superiority over the visual 'cut' as a mediator of information to primary school children has been discussed by Hordley (1977). The present experiment was therefore designed to establish whether zooming in and out may hold separate connotations as hypothesised.

Procedure

A lecture extract was scripted featuring two alternative views on a fictitious aspect of literary criticism. It was then performed for TV twice, by a lecturer seen beside a classroom blackboard. In the second performance the order of the two points was reversed; in all other respects the performances were, as far as possible, identical. During both the performer read his text from an autocue device placed between the lenses of two immediately adjacent cameras, and he appeared to address both cameras directly: the angle displacement between the cameras was imperceptible. Each framed the performer in a basic mid shot though zoomed either in (to close up) or out (to long shot) during the second point in the text. Two simultaneous thirty second recordings of each performance were made possible on this basis; and the whole forms a two way design in which order of points is balanced while the directional effect of zooming, if any, on their perceived importance is observed.

At least ten student subjects viewed each of the four conditions (N = 50 in all), and they were then asked: 'Which of the lecturer's two points, if either, did you feel to be the more important (the first or the second)? And why?'

Results

The responses to the two performances were pooled and the number of preferences for each textual point in the separate 'zoom' conditions was calculated. Via a chi-squared test no significant interaction between zooming and perceived importance was observed. Only one of the 50 subjects attributed her choice of points to the specific connotations of the zoom effect hypothesised.

Discussion

Connotations of the zoom effect regarding verbal significance were not apparent to students as a whole, and the experimental hypothesis was shown to be tenable by the responses of one of them only.

The coded effects of zooming previously hypothesised are clearly too rarified to be apparent intuitively to student subjects in general. But if *just one of them* affirms that the hypothesis is tenable, the likelihood that other subjects may be taught the code subsequently is increased. If both the producers and viewers of TV material recognise the symbolic potential of particular image variations, however esoteric it may initially seem, they may gradually come to use it for reliable communication between them. Of course, the device must be consistently used for the one effect at a given time and its usage must be instantly recognisable: this principle has been exemplified by Webster and Cox (1974) in a study of colour coding in TV captions. From an experiment which may at first appear null and void we may therefore derive specific suggestions for future testing.

For let us be moderate in the approach we adopt to such problems. As indicated above an excessively rigid adherence to *a priori* criteria for either experimental design, analysis or the interpretation of results, may be as ineffectual in the investigation of complex matters as theoretical speculations that lack all discipline. As new criteria for a meaningful result become apparent the methodology should be modified to take stock of them. In studies of an applied problem as vexed as the effects of TV, the most acceptable approach is not necessarily that which adheres to rigorous textbook precepts; it is that which, regardless of method, indicates original uses capable of being tested in their own right.

Communication and aesthetics

We began this report by regretting that so little is known about the psychological processes underlying media communication, about the potential effects of media technique, and about the individual audience differences that shape them; and in the subsequent chapters we have aimed to measure a few aspects of each. A suitable epitaph to the work is provided by Plato: 'How hard, or rather impossible, is the attainment of certainty about questions such as these in the present life. And yet he would be a poor creature who . . . abandoned the task before he had examined them from every side and could do no more!' There is certainly plenty more that scientists and practitioners can do to develop our understanding and usage of media henceforward, and much scope for collaboration between them in the process. But many other questions are raised during such an attempt which may momentarily distract each of them. All media have an aesthetic function which neither science nor logic may ever fully explain; but its very complexity can provide an irresistible challenge. The problems that challenged Plato were less mundane than the effects of TV—or perhaps they were simply less applied. They certainly concerned communication and the effects of information on human response. They also raised philosophical questions of truth and beauty and Man's mediation of them via the 'woven web of guesses' since referred to by Popper (1963). Each of these issues bewilders us to the same extent today as it has done in the past; and, for want of a fully capable methodology, the study of Man's aesthetic sensitivity remains at the fringes of psychology as the province of romantics and zealots. Yet the phenomena of communication and aesthetics are potentially one and the same, referring to interactions between attitude complexity, individual difference and mediating technique in contexts such as described above. If we can begin to unravel the patterns they describe in one context, perhaps we can learn to do so in the others simultaneously.

Thus, as the images of TV may be seen to produce aesthetic effects upon reactions to news and current affairs, so may works of music, art and literature be seen to convey definite information to those who examine them. Whether art is a reliable means of communication, or communicated messages a simple artifice without meaning, are both questions necessitating comparisons between intention and interpretation by methods such as we have employed in the present study. The study developed in part from an interest in the phenomenon of synaesthesia, the capacity to perceive music, for instance, in terms of colour (Baggaley, 1972); and Osgood's original design of the semantic differential technique arose from the same interest (Karwoski et al., 1942) and a wish to measure the intensity of music-colour associations in simple scalar terms. The studies of meaning measurement reported by Osgood et al. (1957) actually follow, in a highly logical manner, from semiotic principles now gaining the attention of media researchers in their raw and less yielding forms twenty years later; and it is well to remember these ready made connections when formulating semiotic hypotheses concerning aesthetic processes henceforward.

For the processes whereby musical signifiers are vested with visual significance, and vice versa, are no different in conception from those whereby camera angle may connote credibility or the use of lecture notes be construed in relation to expertise. The semiotic problem underlying both phenomena is fixed and constant, remaining so whether it is investigated by the techniques of linguistic analysis, signal detection theory or, as here, mental measurement. It will be interesting to see whether methods developed on these individual theoretical bases in future will help to render our understanding of the mediation processes underlying all human experience more explicit.

Bibliography

Altheide, D.L. (1976), *Creating Reality: How TV News Distorts Events,* Beverly Hills, Sage.

Anderson, K. (1961), *An Experimental Study of the Interaction of Artistic and Non-artistic Ethos in Persuasion,* Doctoral dissertation, University of Wisconsin.

Argyle, M. (1969), *Social Interaction,* London, Methuen.

Argyle, M. (1975), *Bodily Communication,* London, University Paperbacks.

Argyle, M. and McHenry, R. (1971), 'Do spectacles really affect judgements of intelligence?', *Brit.J.Soc.Clin.Psychol.,* vol.10.

Arnheim, R. (1969), *Visual Thinking,* Berkeley, University of California Press.

Aronson, E. (1973), *The Social Animal,* London, Freeman.

Aronson, E. and Carlsmith, J.M. (1968), 'Experimentation in social psychology', in Lindzey, G. and Aronson, E. (eds.), *Handbook of Social Psychology,* vol.2, Reading, Mass., Addison-Wesley.

Attneave, F. (1959), *Applications of Information Theory to Psychology,* New York, Holt Rinehart.

Aylward, T.J. (1960), 'A study of the effect of production technique on a televised lecture', *Dissertation Abstracts,* vol.21.

Baggaley, J.P. (1972), *Colour & Musical Pitch,* Ph.D. Thesis: University of Sheffield.

Baggaley, J.P. (1973), 'Developing an effective educational medium', *Prog. Learning & Educ. Technol.,* vol.10.

Baggaley, J.P. (1977), *The Psychological and Educational Impact of TV Production,* London, Social Science Research Council Report.

Baggaley, J.P. (1978), 'Television codes and audience response', *J.Educ.TV,* vol.4; also British Film Institute Educational Advisory Document.

Baggaley, J.P. and Duck, S.W. (1975), 'Communication effectiveness in the educational media', in Baggaley, J.P., Jamieson, G.H. and Marchant, H. (eds.) *Aspects of Educational Technology,* vol.8, Bath, Pitman.

Baggaley, J.P. and Duck, S.W. (1976), *Dynamics of Television,* Farnborough, Saxon House, and Lexington, Lexington Books.

Baggaley, J.P. and Duck, S.W. (1979), 'On making charitable appeals more appealing', *J.Educ.TV.,* vol.5.

Barrington, H. (1972), 'Instruction by television: two presentations compared', *Educ.Research,* vol.14.

Bateson, G. (1973), *Steps to an Ecology of Mind,* St Albans, Herts., Paladin.

Beach, D.E. (1960), *An Analysis of the Retention Involved in Three Methods of Television News Presentation*, MA Thesis, University of Ohio.

Berlo, D.K. (1961), *An Empirical Test of a General Construct of Credibility*, Paper presented at SAA Convention, New York City.

Bernard, L.L. (1926), *Introduction to Social Psychology*, New York, Holt.

Birt, J. and Jay, P. (1975/6), Articles in *The Times*, 28 February 1975, 30 September 1975, 2-3 September 1976.

Blumler, J. and Katz, E. (eds.) (1974), *The Uses of Mass Communications*, Beverly Hills, Sage.

Blumler, J. and McQuail, D. (1968), *Television in Politics: Its Uses and Influences*, London, Faber & Faber.

Booth, A. (1970-71), 'The recall of news items', *Public Opinion Quarterly*, vol.34.

Brandon, J.R. (1956), 'An experimental television study: the relative effectiveness of presenting factual information by the lecture, interview and discussion methods', *Speech Monographs*, vol.23.

Broadbent, D.E. (1958), *Perception and Communication*, London, Pergamon.

Brown, R. (1965), *Social Psychology*, New York, The Free Press.

Bruning, J.L. and Kintz, B.L. (1968), *Computational Handbook of Statistics*, Glenview, Illinois, Scott Foresman.

Burt, C.L. (1937), 'Correlations between persons', *Brit.J.Psychol.*, vol.28.

Buscombe, E. (ed.) (1975), *Football on Television*, London, British Film Institute TV Monograph 4.

Calkins, D.R. (1971), 'Cognitive dissonance: effect on adult learning when incongruous instructors teach through the medium of television', *Dissertation Abstracts International*, vol.32.

Canter, D., West, S. and Wools, R. (1974), 'Judgements of people and their rooms', *Brit.J.Soc.Clin.Psychol.*, vol.13.

Cattell, R.B. (1949), 'r_p and other coefficients of pattern similarity', *Psychometrika*, vol.14.

Cattell, R.B. (1966a), 'The scree test for the number of factors', *Multivariate Behav.Research*, vol.1.

Cattell, R.B. (1966b), *Handbook of MultivariateExperimental Psychology*, Chicago, Rand-McNally.

Chu, G. and Schramm, W. (1967), *Learning from Television: What the Research Says*, Washington, National Association of Educational Broadcasters.

Clark, R.E. (1975), 'Constructing a taxonomy of media attributes for research purposes', *AV Communication Review*, vol.23.

Clark, R.E. (1978), 'Visual instruction and visual aptitudes', in

Sullivan, A. (ed.), Proceedings of First International Conference on *Experimental Research in Videotaped Instruction*, Newfoundland, Memorial University.
Clark, R.E. and Snow, R.E. (1975), 'Alternative designs for instructional technology research', *AV Communication Review*, vol.23.
Cobin, M.T. and McIntyre, C.J. (1961), *The Development and Application of a New Method to Test the Relative Effectiveness of Specifical Visual Production Techniques for Instructional Television*, Urbana, University of Illinois.
Coldevin, G.O. (1975a), 'The differential effects of voice-over, superimposition and combined review treatments as production strategies for ETV programming', in Baggaley, J.P., Jamieson, G.H. and Marchant, H. (eds.) *Aspects of Educational Technology*, vol.8, London, Pitman.
Coldevin, G.O. (1975b), 'Spaced, massed and summary treatments, as review strategies for ITV production', *AV Communication Review*, vol.23.
Coldevin, G.O. (1976), 'Comparative effectiveness of ETV presentation variables', *J.Educ.TV*, vol.2.
Coldevin, G.O. (1977), 'Factors in ETV presenter selection: effects of stereotyping', *Brit.J.Educ.Technol.*, vol.8.
Coldevin, G.O. (1978a,b), 'Experiments in TV presentation strategies' (Parts I and II), *Educ.Broadcasting International*, vol.11, nos.1 and 3.
Coldevin, G.O. (1979), *Experimental Research in TV Production Techniques: Current Directions*, Paper to Annual Conference of the Educational TV Association, University of York.
Collins, R. (1976), *Television News*, London, British Film Institute TV Monograph 5.
Colvin, S.S. and Bagley, W.C. (1915), *Human Behavior: A First Book in Psychology for Teachers*, London, MacMillan.
Comrey, A.L. (1973), *A First Course in Factor Analysis*, New York, Academic Press.
Connolly, C.P. (1962), *An Experimental Investigation of Eye-Contact on Television*, MA Thesis, University of Ohio.
Corner, J. (1978), 'Where students part company', *Media Reporter*, vol.2.
Cronbach, L.J. (1956), *Essentials of Psychological Testing*, New York, Harper & Row.
Cronbach, L.J. (1957), 'The two disciplines of scientific psychology', *Amer.Psychologist*, vol.12.
Cronbach, L.J., Rajaratnam, N. and Glesser, G.C. (1963), 'Theory of generalizability: a liberation of reliability theory', *Brit.J.Statist. Psychol.*, vol.16.
Cronbach, L.J. and Snow, R.E. (1977), *Aptitudes and Instructional Methods*, New York, Irvington Press.

Dannheisser, P. (1975), 'A closer look at the audience', *Independent Broadcasting*, vol.3.
Davis, L.K. (1978), 'Camera eye-contact by the candidates in the Presidential debates of 1976', *Journalism Quarterly*, vol.55.
Duck, S.W. (1973), *Personal Relationships and Personal Constructs*, London, Wiley.
Dyer, R. (1973), *Light Entertainment*, London, British Film Institute TV Monograph 2.

Edwards, A.C. (1965), *Experimental Design in Psychological Research*, London, Holt, Rinehart and Winston.
Edwardson, M., Grooms, D. and Pringle, P. (1976), 'Visualization and TV news information gain', *J.Broadcasting*, vol.20.
Eisenstein, S.M. (1947), *The Film Sense*, New York, Harcourt and Brace.
Ekman, P. and Friesen, W.V. (1969), 'Non-verbal leakage and clues to deception', *Psychiatry*, vol.32.
Ellery, J.B. (1959), *A Pilot Study of the Nature of Aesthetic Experiences Associated with Television and its Place in Education*, Detroit, Wayne State University.
Eysenck, H.J. (1953), *The Structure of Human Personality*, London, Methuen.
Eysenck, H.J. (1978), 'Testing TV violence', letter to *The Observer*, 1 October.
Eysenck, H.J. and Nias, D.K. (1978), *Sex. Violence and the Media*, London, Temple Smith.

Festinger, L. (1954), 'The Theory of social comparison processes', *Human Relations*, vol.7.
Festinger, L. (1957), *A Theory of Cognitive Dissonance*, New York, Harper and Row.
Findahl, O. and Hoijer, B. (1972), *Man as Receiver of Information*, Stockholm, Swedish Broadcasting Corporation.
Findahl, O. and Hoijer, B. (1976), *Fragments of Reality*, Stockholm, Swedish Broadcasting Corporation.
Findahl, O. and Hoijer, B. (1977), *How Important is Presentation?* Stockholm, Swedish Broadcasting Corporation.
Fiske, J. and Hartley, J. (1978), *Reading Television*, London, Methuen.

Garnham, N. (1973), *Structures of Television*, London, British Film Institute TV Monograph 1.
Garrett, H.E. (1964), *Statistics in Psychology and Education*, New York, Longman.
Glasgow University Media Group (1976), *Bad News*, vol.1, London, Routledge and Kegan Paul.

Greimas, A.J. (1966), *La Semantique Structurale,* Paris, Larousse.
Guiraud, P. (1975), *Semiology,* London, Routledge and Kegan Paul.
Gunter, B. (1979), 'The effect of pictures upon memory for brief television news items', *J.Educ.TV,* vol.5.

Harman, H.H. (1967), *Modern Factor Analysis,* Chicago, University of Chicago Press.
Hartley, J., Holt, J. and Swain, F. (1970), 'The effects of pre-tests, interim tests, and age on post-test performance following self-instruction', *Prog.Learning and Educ.Technol.,* vol.7.
Hawkes, T. (1977), *Structuralism and Semiotics,* London, Methuen.
Hazard, W.R. (1962-63), 'On the impact of television's pictured news', *J.Broadcasting,* vol.7.
Heider, F. (1958), *The Psychology of Interpersonal Relations,* New York, Wiley.
Hocking, J.E., Margreiter, D.G. and Hylton, C. (1975), *Intra-audience Effects: A Field Test,* Paper presented to the International Communication Association Convention, Chicago.
Hofmann, R.J. (1975), 'On the proportionate contributions of transformed factors to common variances', *Multivariate Behav.Research,* vol.10.
Hood, S. (1975), 'Visual literacy examined', in Luckham, B. (ed.) *Proceedings of the Sixth Symposium on Broadcasting Policy—'Audio-Visual Literacy',* University of Manchester.
Hordley, R. (1977), 'Specific factors in the perception of film and television', *J.Educ.TV,* vol.3.
Hovland, C.I., Janis, I.L. and Kelley, H.H. (1953), *Communication and Persuasion,* New Haven, Yale University Press.
Hovland, C.I. and Weiss, W. (1952), 'The influence of source credibility on communication effectiveness', *Public Opinion Quarterly,* vol.15.

Jacquinot, G. (1977), *Image et Pedagogre,* Paris, L'Universite de Paris.
Jamieson, G.H. (1973), 'Visual media in a conceptual framework for the acquisition of knowledge', *Prog.Learning and Educ.Technol.,* vol.10.
Jamieson, G.H., Thompson, J.O. and Baggaley, J.P. (1976), 'Intention and interpretation in the study of communication', *J.Educ.TV,* vol.2.

Kaiser, H.F. (1956), *The Varimax Method of Factor Analysis,* Doctoral dissertation, University of California.
Kaiser, H.F. (1960), 'The application of electronic computers in factor analysis', *Educ.Psychol.Measurement,* vol.20.
Kaiser, H.F. and Caffrey, J. (1965), 'Alpha Factor Analysis', *Psychometrika,* vol.30.

Kaplan, K.J. (1972), 'On the ambivalence–indifference problem in attitude theory and measurement: a suggested modification of the semantic differential technique', *Psychol.Bull.*, vol.77.
Karwoski, T., Odbert, H. and Osgood, C.E. (1942), 'Studies in synesthetic thinking: II. The role of form in visual responses to music', *J.Gen.Psychol.*, vol.26.
Katz, E., Adoni, E. and Parness, P. (1977), 'Remembering the news: what the picture adds to recall', *Journalism Quarterley*, vol.54.
Katz, E., Blumler, J.G. and Gurevitch, M. (1974), in Blumler, J.G. and Katz, E., *The Uses of Mass Communications*, London, Sage.
Kay, H., Dodd, B. and Sime, M. (1968), *Teaching Machines and Programmed Instruction*, Harmondsworth, Penguin.
Kelley, H.H. (1967), in Levine, D. (ed.) *Nebraska Symposium on Motivation*, Lincoln, Nebraska University Press.
Kelley, H.H. and Woodruff, C.L. (1956), 'Members' reactions to apparent group approval of a counter-norm communication', *J.Abn.Soc. Psychol.*, vol.52.
Kelly, G.A. (1955), *The Psychology of Personal Constructs*, New York, Norton.
Kelman, H.C. and Hovland, C.I. (1953), ' "Reinstatement" of the communicator in delayed measurement of opinion change', *J.Abn.Soc. Psychol.*, vol.48.
Kendall, M. and Stuart, A. (1961), *The Advanced Theory of Statistics*, vol.II, New York, Hafner.
Kirkpatrick, E.A. (1916), *Fundamentals of Child Study*, New York, MacMillan.
Kreimer, O. (1974), *Open Sesame: A Key to the Meaning of the Educational Broadcast Message*, Stanford University, Institute for Communication Research.
Krull, R., Watt, J.H. and Lichty, L.W. (1977), 'Entropy and structure: two measures of complexity in television programs', *Communication Research*, vol.4.

Landy, D. (1972), 'The effects of an overhead audience's reaction and attractiveness on opinion change', *J.Exp.Soc.Psychol.*, vol.8.
Levy, L.H. (1960), 'Context effects in social perception', *J.Abn.Soc. Psychol.*, vol.61.
Lometti, G.E., Reeves, B. and Bybee, C.R. (1977), 'Investigating the assumptions of uses and gratifications research', *Communication Research*, vol.4.

McCain, T.A., Chilberg, J. and Wakshlag, J. (1977), 'The effect of camera angle on source credibility and attraction', *J.Broadcasting*, vol.21.

McCroskey, J.C. (1966), 'Scales for the measurement of ethos', *Speech Monographs*, vol.33.

McDaniel, D.O. (1974), *TV Newsfilm: A Study in Audience Perception*, Ann Arbor, Michigan, University Microfilms Inc.

McDougall, W. (1908), *An Introduction to Social Psychology*, London, Methuen.

McGuire, W. (1972), In McLintock, C.G. (ed.) *Experimental Social Psychology*, New York, Rinehart and Winston.

McLuhan, M. (1964), *Understanding Media*, London, Routledge and Kegan Paul.

McMenamin, M.J. (1974), 'Effect of instructional television on personality perception', *A V Communication Review*, vol.22.

Mandell, L.M. and Shaw, D.L. (1973), 'Judging people in the news unconsciously: effect of camera angle and bodily activity', *J.Broadcasting*, vol.17.

Markham, D. (1968), 'The dimensions of source credibility of television newscasters', *J. Communication*, vol.18.

Masterman, L. (1978), 'TV specificity and participatory planning: two participatory exercises', *J.Educ.TV*, vol.4.

Messaris, P., Eckman, B. and Gumpert, G. (1979), 'Editing structure in the televised versions of the 1976 Presidential debates', *J.Broadcasting*, vol.23.

Metz, C. (1974), *Film Language: A Semiotics of the Cinema*, New York, Oxford University Press.

Mielke, K.W. (1970), 'Media-message interactions in TV', *Viewpoints*, vol.46, University of Indiana.

Miller, J. (1971), *McLuhan*, London, Fontana.

Miller, W. (1969), 'Film movement and affective response and the effect on learning and attitude formation', *A V Communication Review*, vol.17.

Morgan, F. (1978), *Using Television*, IET Broadcast Evaluation Report No.27, Milton Keynes, Open University.

Muldoon, J.F. and Ray, O.S. (1958), 'A comparison of pattern similarity as measured by six statistical techniques and eleven clinicians', *Educ. and Psychol.Measurement*, vol.18.

Nie, N.H., Hull, C.H., Jenkins, J.G., Steinbrenner, K. and Bent, D.H. (1975), *Statistical Package for the Social Sciences*, New York, McGraw-Hill.

Osgood, C.E., Suci, G. and Tannenbaum, P. (1957), *The Measurement of Meaning*, Chicago, University of Illinois Press.

Packard, V. (1964), *The Hidden Persuaders*, Harmondsworth, Penguin.

Palmer, E.L. (1969), 'Research at the Children's Television Workshop', *Educ.Broadcasting Review*, vol.3.
Pateman, T. (1974), *Television and the February 1974 General Election*, London, British Film Institute TV Monograph 3.
Perry, R. and Williams, R. (1978), 'Dr Fox's Lecture paradigm', in Sullivan, A. (ed.), Proceedings of First International Conference on *Experimental Research in Videotaped Instruction*, Newfoundland, Memorial University.
Popper, K. (1963), *Conjectures and Refutations: The Growth of Scientific Knowledge*, London, Routledge and Kegan Paul.
Postman, L. and Egan, J.P. (1949), *Experimental Psychology*, New York, Harper.
Pudovkin, V.I. (1958), *Film Technique and Film Acting*, London, Vision Press.

Rowley, C. (1975), 'ITV's programme balance: 1970-75', *Independent Broadcasting*, vol.6.
Rummel, R.J. (1967), 'Understanding factor analysis', *Conflict Resolution*, vol.11.
Rummel, R.J. (1970), *Applied Factor Analysis*, Evanston, Northwestern University Press.

Salomon, G. (1972), 'Can we affect cognitive skills through visual media?', *AV Communication Review*, vol.20.
Salomon, G. and Cohen, A. (1977), 'Television formats, mastery of mental skills, and the acquisition of knowledge', *J.Educ.Psychol.*, vol.69.
Saussure, F. de (1915), *Cours de Linguistique Generale*, Lausanne/Paris, Bally and Sechehaye, (Translation: London, Fontana 1974).
Scheff, T.J. (1973), 'Intersubjectivity and emotion', *Amer.Behavioral Scientist*, vol.16.
Schegloff, E.A. (1968), 'Sequencing in conversational openings', *Amer. Anthropologist*, vol.70.
Schlater, R. (1970), 'Effect of speed of presentation on recall of television messages', *J.Broadcasting*, vol.14.
Schramm, W. (1971), *The Research on Content Variables in ITV*, Stanford University, Institute for Communication Research.
Schramm, W. (1975), *Big Media, Little Media*, Beverly Hills, Sage.
Sells, P. (1979), *Effects of TV Presentation Upon Attributed Characteristics of a Performer*, Unpublished manuscript, Centre for Communication Studies, University of Liverpool.
Severin, W. (1967), 'The effectiveness of relevant pictures in multichannel communications', *AV Communication Review*, vol.15.
Shannon, C.E. and Weaver, W. (1949), *The Mathematical Theory of*

Communication, Urbana, University of Illinois Press.
Sherrington, R. (1978), Review: 'Dynamics of Television', *Educ.Broadcasting International,* vol.11.
Shulman, M. (1975), *The Ravenous Eye,* London, Coronet.
Siegel, S. (1956), *Nonparametric Statistics for the Behavioral Sciences,* London, McGraw-Hill.
Skinner, B.F. (1938), *The Behaviour of Organisms,* New York, Appleton-Century.
Slater, P. (1969), 'Theory and technique of the repertory grid', *Brit.J. Psychiat.,* vol.115.
Smith, A. (1978), 'Censoring violence on TV isn't so simple', *The Observer,* 24 September.
Snider, J.G. and Osgood, C.E. (1969), *Semantic Differential Technique,* Chicago, Aldine.
Stephenson, W. (1953), *The Study of Behaviour,* Chicago, University of Chicago Press.
Suchett-Kaye, C. (1972), 'Personality factors and self-instruction: a survey', *Prog.Learning and Educ.Technol.,* vol.9.
Sullivan, A., Andrews, E.A., Hollingshurst, F., Maddigan, R. and Noseworthy, C.M. (1977), 'The relative effectiveness of instructional television', *J.Educ.TV,* vol.3.
Sullivan, A., Andrews, E.A., Maddigan, R. and Noseworthy, C.M. (1978), 'Learning from television, I', in Sullivan, A. (ed.) Proceedings of First International Conference on *Experimental Research in Videotaped Instruction,* Newfoundland, Memorial University.
Sullivan, A., Andrews, E.A., Maddigan, R. and Noseworthy, C.M. (1979), 'Learning from television, II', in Sullivan, A. (ed.) Proceedings of Second International Conference on *Experimental Research in Videotaped Instruction,* Newfoundland, Memorial University.
Swets, J.A. (ed.) (1964), *Signal Detection and Recognition by Human Observers,* New York, Wiley.

Tannenbaum, P.H. (1955), 'What effect when TV covers a congressional hearing?', *Journalism Quarterly,* vol.32.
Tannenbaum, P.H. (1956), 'The effect of background music on interpretation of stage and television drama', *AV Communication Review,* vol.4.
Thurstone, L.L. (1947), *Multiple Factor Analysis,* Chicago, University of Chicago Press.
Travers, R.M. (1964, 1966), *Research and Theory Related to Audio-Visual Information Transmission,* Bureau of Educational Research, University of Utah.

Vaughan, D. (1976), *Television Documentary Usage,* London, British Film Institute TV Monograph 6.

Wakshlag, J.J. (1979), 'Sequential structure in televised behaviour', *J.Broadcasting,* vol.23.

Warr, P. and Knapper, C. (1968), *The Perception of People and Events,* London, Wiley.

Warren, H.C. (1919), *Human Psychology,* London, Constable.

Webster, B.R. and Cox, S.M. (1974), 'The value of colour in educational television', *Brit.J.Educ.Technol.,* vol.5.

Westley, B.H. and Mobius, J.B. (1960), *The Effects of 'Eye-Contact' in Televised Instruction,* Madison, University of Wisconsin TV Laboratory.

Williams, F. (1963), 'A factor analysis of judgements of radio broadcasters', *J.Broadcasting,* vol.7.

Williams, R. (1973), *Cultural Studies and Communication,* Paper presented to International Conference on the Future of Communication Studies, Heathrow, London.

Williams, R. (1974), *Television: Technology and Cultural Form,* London, Fontana.

Winer, B.J. (1971), *Statistical Principles in Experimental Design,* London, McGraw-Hill.

Wollen, P. (1974), *Signs and Meaning in the Cinema,* London, Secker and Warburg.

Woodworth, R.S. (1926), *Psychology: A Study of Mental Life,* London, Methuen.

Wurtzel, A.H. and Dominick, J.R. (1971-72), 'Evaluation of television drama: interaction of acting styles and shot selection', *J.Broadcasting,* vol.16.

Index

THE AUTHOR

Jon Baggaley gained first and second degrees in psychology at the University of Sheffield, and has been Lecturer in Communication Studies at the University of Liverpool since 1971. He has written and edited several previous volumes concerning the educational media and is co-author of *Dynamics of Television* with Steven Duck. From 1979 he is appointed Associated Professor in the Institute for Research in Human Abilities at Memorial University of Newfoundland. He also directs and performs in musical theatre.

Margaret Ferguson and *Philip Brooks* have assisted Dr Baggaley in the experimental work reported in the present book, supported financially by the Social Science Research Council. Having graduated in psychology from Strathclyde University and Plymouth Polytechnic respectively, both have a particular interest in the phenomena of perception and aesthetics. Philip Brooks is currently completing his doctorate in the psychology of thinking.